D1346010

SCOPE 34

Practitioner's Handbook on the Modelling of Dynamic Change in Ecosystems

SCOPE 34

Practitioner's Handbook on the Modelling of Dynamic Change in Ecosystems

J. N. R. Jeffers
Institute of Terrestrial Ecology
Grange-over-Sands, Cumbria

Published on behalf of the
Scientific Committee on Problems of the Environment (SCOPE)
of the
International Council of Scientific Unions (ICSU)
by

JOHN WILEY & SONS
Chichester · New York · Brisbane · Toronto · Singapore

Library of Congress Cataloging-in-Publication Data:

Jeffers, J. N. R. (John Norman Richard)
Practitioner's handbook on the modelling of dynamic change in ecosystems/
J. N. R. Jeffers.
p. cm. — (SCOPE; 34)
1. Ecology — Simulation methods — Handbooks, manuals, etc.
I. International Council of Scientific Unions. Scientific Committee on Problems of the Environment. II. Title. III. Series: Scope (Series); 34.
QH541.15.S5J44 1988
574.5′0724 — dc19 87–29555
ISBN 0-471-10519-8:

British Library Cataloguing in Publication Data:

Jeffers, J. N. R.
Practitioners handbook on the modelling of dynamic change in ecosystems. — (SCOPE; 34).
1. Ecology — Mathematical models
I. Title II. International Council of Scientific Unions. *Scientific Committee on Problems of the Environment*
III. Series
574.5′0724 QH541.15.M3

ISBN 0 471 10519 8

Typeset by Mathematical Composition Setters Ltd
7 Ivy Street, Salisbury, Wiltshire, England, SP1 2AY
Printed and bound in Great Britain by Biddles Ltd., Guildford

SCOPE 1: Global Environmental Monitoring 1971, 68pp (out of print)

SCOPE 2: Man-Made Lakes as Modified Ecosystems, 1972, 76pp

SCOPE 3: Global Environmental Monitoring Systems (GEMS): Action Plan for Phase 1, 1973, 132pp

SCOPE 4: Environmental Sciences in Developing Countries, 1974, 72pp

Environment and Development, proceedings of SCOPE/UNEP Symposium on Environmental Sciences in Developing Countries, Nairobi, February 11–23, 1974, 418pp.

SCOPE 5: Environmental Impact Assessment: Principles and Procedures, Second Edition, 1979, 208pp

SCOPE 6: Environmental Pollutants: Selected Analytical Methods, 1975, 277pp (available from Butterworth & Co. (Publishers) Ltd, Sevenoaks, Kent, England

SCOPE 7: Nitrogen, Phosphorus, and Sulphur: Global Cycles, 1975, 192pp (available from Dr Thomas Rosswall, Swedish Natural Science Research Council, Stockholm, Sweden)

SCOPE 8: Risk Assessment of Environmental Hazard, 1978, 132pp

SCOPE 9: Simulation Modelling of Environmental Problems, 1978, 128pp

SCOPE 10: Environmental Issues, 1977, 242pp

SCOPE 11: Shelter Provision in Developing Countries, 1978. 112pp

SCOPE 12: Principles of Ecotoxicology, 1978, 372pp

SCOPE 13: The Global Carbon Cycle, 1979, 491pp

SCOPE 14: Saharan Dust: Mobilization, Transport, Deposition, 1979, 320pp

SCOPE 15: Environmental Risk Assessment, 1980, 176pp

SCOPE 16: Carbon Cycle Modelling, 1981, 404pp

SCOPE 17: Some Perspectives of the Major Biogeochemical Cycles, 1981, 175pp

SCOPE 18: The Role of Fire in Northern Circumpolar Ecosystems, 1983, 344pp

SCOPE 19: The Global Biogeochemical Sulphur Cycle, 1983, 495pp

SCOPE 20: Methods for Assessing the Effects of Chemicals on Reproductive Functions, 1983, 568pp

SCOPE 21: The Major Biogeochemical Cycles and Their Interactions, 1983, 554pp

SCOPE 22: Effects of Pollutants at the Ecosystem Level, 1984, 443pp

SCOPE 23: The Role of Terrestrial Vegetation in the Global Carbon Cycle: Measurement by Remote Sensing, 1984, 272pp

SCOPE 24: Noise Pollution, 1986, 472pp

SCOPE 25: Appraisal of Tests to Predict the Environmental Behaviour of Chemicals, 1985, 400pp

SCOPE 26: Methods for Estimating Risks of Chemical Injury: Human and Non-human Biota, 1985, 712pp

SCOPE 27: Climate Impact Assessment: Studies of the Interaction of Climate and Society, 1985, 649pp

SCOPE 28: Environmental Consequence of Nuclear War
Volume I Physical, 1985, 342pp
Volume II Ecological and Agricultural Effects, 1985, 523pp

SCOPE 29: The Greenhouse Effect, Climate Change and Ecosystems, 1986, 574pp

SCOPE 30: Methods for Assessing the Effects of Mixtures of Chemicals, 1987, 928pp

SCOPE 31: Occurrence and Pathways of Lead Mercury Cadmium and Arsenic in the Environment, 1987, 384pp

SCOPE 32: Land Transformation in Agriculture, 1987, 384pp

SCOPE 33: Nitrogen Cycling in Coastal Marine Environments, 1988, 480pp

SCOPE 34: Practitioner's Handbook on the Modelling of Dynamic Change in Ecosystems, 1988, 176pp

SCOPE 35: Scales and Global Change: Spatial and Temporal Variability in Biospheric and Geospheric Processes, 1988

SCOPE 36: Acidification in Tropical Countries, 1988

Funds to meet SCOPE expenses are provided by contributions from SCOPE National Committees, an annual subvention from ICSU (and through ICSU, from UNESCO), an annual subvention from French Ministère de l'Environment et du Cadre de Vie, contracts with UN Bodies, particularly UNEP, and grants from Foundations and industrial enterprises.

International Council of Scientific Unions (ICSU)
Scientific Committee on the Problems of the Environment (SCOPE)

SCOPE is one of a number of committees established by a non-governmental group of scientific organizations, the International Council of Scientific Unions (ICSU). The membership of ICSU includes representatives from 68 National Academies of Science, 18 International Unions and 12 other bodies called Scientific Associates. To cover multidisciplinary activities which include the interest of several unions, ICSU has established 10 scientific committees, of which SCOPE is one. Currently, representatives of 34 member countries and 15 Unions and Scientific Committees participate in the work of SCOPE, which directs particular attention to the needs of developing countries. SCOPE was established in 1969 in response to the environmental concerns emerging at that time; ICSU recognized that many of these concerns required scientific inputs spanning several disciplines and ICSU Unions. SCOPE's first task was to prepare a report on Global Environmental Monitoring (SCOPE 1, 1971) for the UN Stockholm Conference on the Human Environment.

The mandate of SCOPE is to assemble, review, and assess the information available on man-made environmental changes and the effects of these changes on man; to assess and evaluate the methodologies of measurement of environmental parameters; to provide an intelligence service on current research; and by the recruitment of the best available scientific information and constructive thinking to establish itself as a corpus of informed advice for the benefit of centres of fundamental research and of organizations and agencies operationally engaged in studies of the environment.

SCOPE is governed by a General Assembly, which meets every three years. Between such meetings its activities are directed by the Executive Committee.

R. E. Munn
Editor-in-Chief
SCOPE Publications

Executive Secretary: V. Plocq

Secretariat: 51 Bld de Montmorency
75016 PARIS

Contents

Foreword

There exists a wide range in the impact of man's actions on natural ecosystems. When impact is limited, the ecosystem tends to retain most of its original attributes. When impact is very severe, the natural system is often replaced. In the former category are many areas such as national parks and equivalent reserves, or lands where traditional forest and range management techniques are applied and where there is usually a low energy supplement associated with human activity. In the latter category are intensively used rural and urban lands, in which there is usually a high energy supplement. Between the two extremes is a range of increasing habitat modification usually associated with an increasing amount of energy input and a decreasing dependence on the natural environment and biota.

Broadly speaking, it appears that as the degree of habitat modification increases and as natural communities are altered or replaced by communities composed largely of exotic species, a knowledge of the structural and functional relationships of the original ecosystems becomes less relevant to the management of the modified systems. However, the reverse also appears to be true, and in systems that are utilized essentially as managed natural systems, a knowledge of structural and functional relationships probably provides the soundest basis for effective long-term management, consistent with both conservation and maintenance of productivity. These systems include many terrestrial woodland and forest ecosystems, and many non-woodland systems, such as arid rangeland and alpine and tundra systems, in which environmental conditions are too extreme for forest development.

Ecosystems traditionally utilized for agriculture are being intensively studied by many governmental and nongovernmental organizations, at both national and international levels. By comparison, managed natural systems have received relatively little attention, yet when they are utilized by man they tend to become rapidly destabilized, with associated severe, and sometimes essentially irreversible, changes in their physico-chemical environmental attributes as well as in their biological characteristics.

It was within this context that ICSU's Scientific Committee on Problems of the Environment (SCOPE) and UNESCO's Man and Biosphere (MAB) Programme agreed, in late 1974 to jointly sponsor and convene a technical workshop dealing with dynamic changes in ecological systems, with special reference to contemporary concepts relating to such changes, techniques available for their study, and the degree to which such concepts and techniques could contribute to basic ecological research and to the management of natural and near-natural ecological systems. Professor Ralph Slatyer, of the Department of the Environmental Biology, Australian National University, Canberra, generously agreed to be responsible for the scientific preparation and direction of the workshop, which took place at Santa Barbara, California, from 12 to 16 January 1976. It was attended by the following participants and observers: E. Ames; M. P. Austin; M. Cain; J. B. Connell; G. R. Conway; J. B. Egunjobi; D. W. Goodall; J. M. Hett; R. W. Hilborn; H. S. Horn; J. N. R. Jeffers; R. M. May; C. H. Muller; W. Niering; I. R. Noble; K. L. Reed; M. A. Robinson; H. H. Shugart; R. O. Slatyer; W. Souss; M. B. Usher; R. H. Whittaker.

One written output of the workshop was a short, 30-page report entitled 'Dynamic changes in terrestrial ecosystems: patterns of change, techniques for study and applications to management'. Edited by Ralph Slatyer, this report was published by UNESCO in collaboration with SCOPE in 1977 as MAB Technical Note 4. The Technical Note attempted to review concepts and ideas current at that time regarding environmental succession, to examine the effectiveness of various methods available for modelling and predicting successional patterns, and to consider how these methods can be applied to the management of forests, rangelands and related natural and near-natural ecological systems. The final two substantive sections of the Technical Note provide conclusions and recommendations relating to research on dynamic changes in terrestrial ecosystems and its application to resource use and land management.

The present practitioner's handbook, published as SCOPE 34, comprises another follow-up to the Santa Barbara workshop. It has been prepared in response to one of the specific recommendations of the workshop, which called for the preparation of a practitioner's manual which would outline the techniques available for the modelling of successional processes, the data requirements and limitations of each of the major techniques, and methods of applying them to ecological processes and to the management of semi-natural communities.

Following an agreement between ICSU-SCOPE and UNESCO-MAB, John Jeffers was invited to prepare the handbook. Until his retirement in 1986, John Jeffers was the Director of the Institute of Terrestrial Ecology in the United Kingdom, and has long been interested in the application of systems analysis and other modelling approaches to the organization of ecological research and to the management of natural resources. The Santa Barbara workshop had recommended that the opportunity be taken, in preparing the manual, to increase the biological realism and practical value of the various modelling approaches. Research would clearly be required for the application of existing models and for the development of more effective modelling procedures. The time taken for this research to be undertaken explains in part the period of time that has elapsed between the workshop and the publication of the present handbook.

Franceso di Castri and Malcolm Hadley
Montpellier and Paris, France

CHAPTER 1

Introduction

This Handbook owes its origins to a Workshop jointly sponsored by the United Nations Educational, Scientific and Cultural Organization (UNESCO) Man and the Biosphere (MAB) Programme, and by the Scientific Committee on Problems of the Environment (SCOPE) of the International Council of Scientific Unions (ICSU). The Workshop was concerned with dynamic changes in ecological systems, with special reference to modern concepts relating to such changes, the techniques available for their study, and the degree to which such concepts and techniques could contribute to basic ecological research, and to the management of natural and semi-natural systems.

Five recommendations relating to research on dynamic changes in terrestrial ecosystems, and to the application of the results of such research to problems of resource use and land management, resulted from this Workshop, which was held in Santa Barbara, California, in January 1976 (Slatyer, 1977). These recommendations may be summarized as follows.

(i) Further work on ecological succession should be encouraged, with a view to providing a thorough understanding of successional dynamics, and the interrelationship between successional processes and other community attributes associated with stability, diversity, dominance and resilience. In particular, more information is needed about the biological processes involved in truncated succession, and its associated characteristics.

(ii) Specific studies of succession in a range of ecosystems, and following a variety of different types of perturbation, should be conducted, with a view to providing the data necessary for the construction of comprehensive ecological and management models. Ideally, these studies should be conducted in areas where most of the key ingredients for model construction are already available, and where a team of ecologists and managers could be assembled.

(iii) A practitioner's manual should be prepared to outline the techniques available for the modelling of successional processes, the data requirements and limitations of each of the major techniques, and methods of applying them to ecological processes and to the management of semi-natural communities.

(iv) In association with (ii) and (iii) above, the opportunity should be taken to increase the biological realism and practical value of the various modelling approaches. It was, therefore, envisaged that research would be required both for the application of existing models and for the development of more effective modelling procedures.

(v) Wherever possible, ecologists and managers should be located, geographically and institutionally, where they can interact to develop ecologically sound management strategies for natural, semi-natural and cultivated ecosystems. In turn, greater opportunities need to be provided for the ecologists to interact with specialists in experimental design, data analysis and interpretation, and in model building. Similarly, greater opportunities need to be provided that influence his selection of management goals and constrain his management strategies.

Rapid advances have been made in recent years, in both theory and experiment, on the general level of understanding of ecological succession. In particular, a much better appreciation now appears to exist of the interplay between different successional patterns, the effect of perturbations on the observed patterns, and on the phenomenon of successional truncation in which long-term site dominance is achieved by species which might normally be assumed to be early-successional, sub-dominant species.

The pattern of succession, in terms of the species present on a site, and their relative abundances, at various periods after a disturbance, is basically dependent on the product of two probabilities. The first of these probabilities is that of a propagule being available at the site under consideration, and the second is the probability of it being able to develop to reproductive age and complete its life cycle. Each probability can be affected, to a different degree, by the frequency, intensity, and scale or area of present perturbations. Knowledge of successional pattern can provide a useful basis for developing models of direct relevance to the management of semi-natural communities such as forests, range-lands, and lands set aside as national parks, biosphere reserves and other types of protected ecosystems. Conversely, patterns of natural succession are of very limited relevance to the problems of managing ecosystems in which human intervention has markedly altered the physical or chemical environment and the composition of plant and animal species.

In contrast, managers of semi-natural and cultivated ecosystems need an understanding of the dynamics of the systems they manage in order to predict the effects of various treatments that they apply to improve the yield of harvestable plants or animals. In particular, these managers need to be able to predict the long-term effects of intervention or perturbations on the fertility of the soil, on the dynamics of the interactions between the plants and animals of the system, and between plants, animals of the system, and between plants, animals and the physical environment. It is relatively easy to allow apparent short-term gains to mask long-term harmful effects due to the depletion of essential mineral elements, the accumulation of toxic substances in soils or in organisms, or disturbance of predator–prey relationships that lead to the extinction of either prey or predator. The study of dynamic change is, therefore, of equal importance to the managers of cropped semi-natural and cultivated systems.

In recent years, the application of powerful modelling procedures to ecological processes has stimulated research into ecological succession and into the possibilities of providing a sound ecological basis for the development of management strategies for natural, semi-natural and cropped ecosystems. These techniques have potential application to a wide spectrum of ecological problems and to the management of forests, range-lands and related systems. However, the effective application of ecological knowledge to problems of ecosystem management requires the utilization of a variety of information — ranging from historical records, the research data, to observations obtained by direct management — and also requires a high degree of interaction between the ecologist on the one hand and the manager on the other.

This Handbook has been designed to provide a ready source of reference to both ecologist and manager for the variety of techniques available for the modelling of dynamic change in ecological systems. Because of the different objectives and the variation in availability of data associated with different studies, and because of the characteristics of the techniques, no single approach can generally be recommended. Moreover, because some of the advantages and disadvantages of the major classes of models are complementary, there is generally something to be gained by using a combination of several approaches. As will be stressed later in the Handbook, the modelling of dynamic change needs to be carefully embedded in an overall systems analysis of the problem, and the objectives of the study must be defined with care. Techniques suited to the simulation of ecological processes for the purpose of understanding the processes may be quite inappropriate when applied to problems of direct management, and vice versa.

Alternatively, a technique may also need to be modified because of limitations in the data available. In order to improve both the general level of understanding of dynamic change in ecological systems, and the effectiveness of the methods used to explore them, well-designed surveys and long-term experiments are frequently needed to provide data sets for model generation and testing. Such research involves a recognition that it is the change in ecosystem parameters that needs to be measured rather than the state at which an ecosystem finds itself at any particular time. In the past, ecological research has tended to concentrate on static rather than dynamic models. This Handbook, therefore, ventures into relatively new ground.

CHAPTER 2

Ecological Systems and Their Dynamics

2.1 DEFINITION OF ECOLOGICAL SYSTEMS

As early as 1935, Sir Arthur Tansley suggested the term 'ecosystem' should be used to describe' ... not only the organism complex but also the whole complex of physical factors forming what we call the environment' (Tansley, 1935). This concept of the interaction of living organisms with the physical and chemical factors of their environment has proved to be one of the most important ideas of science, and ecological systems have become the basic units for studies of the interactions between organisms or between organisms and their environment. Various ways of ordering and studying ecological systems have been suggested, dependent on the scale of investigation and the main focus of attention in the research. The role of any organism depends on its place in the ecosystem, and our ability to manipulate or conserve organisms, communities of plants and animals, or whole ecosystems, depends on our understanding of the complex interrelationships between them.

Particular interest has always been expressed in the ways in which organisms combine in communities which are characteristic of a particular type and place, and which reflect past and present land use. Plant communities and associations over the world as a whole can be broadly classified in biomes. Biomes embrace the major vegetation types of the world, and, within broad limits, have a characteristic productivity. Udvardy (1975) has extended this classification into biogeographical realms, biogeographical provinces within the realms, and major biomes or biome complexes. Biome classifications of this kind, combined with hypotheses about the effects of climatic factors, facilitate comparisons of productivity at regional and world scales.

Measurement of changes in biomes provides information on the effects of major influences such as the clearing of forest, desertization, reafforestation, etc., which are usually a combination of man's activities and climate. Within communities, changes in populations of organisms reflect the responses of communities to climate, to modification by man, and to the natural processes of succession through which one complex of organisms gives way to another, leading ultimately to a climax or truncated climax community. Dynamic changes within communities occur at different rates at different localities, so that variations occur from place to place even within the same communities. Spatial and dynamic heterogeneities, therefore, have a marked effect on the patterns of change in ecological systems. It is essential to develop models or analogues of ecological systems which are capable of representing spatial and dynamic heterogeneity so as to avoid assumptions of homogeneity which cannot be substantiated. Changes in populations are measured by comparing numbers of organisms at particular points of time and space, sometimes distinguishing between stages in the development of the organisms, e.g. eggs, larvae, pupae, adults, etc., or between sexes and age classes. In more detailed studies, it may be possible to follow particular marked individuals at various points in time, and so determine patterns of distribution, survival, reproduction, feeding and death.

An alternative to the study of an ecosystem by investigating changes in populations and communities of organisms is the tracing of flows of energy, nutrients and pollutants through the system. Radiant energy from the sun is trapped by green plants and combined with various chemical elements to form organic compounds which enable all the essential properties of life to proceed in living organisms. Such organisms can be classified according to their trophic levels, i.e. by their mode of nutrition. Green plants which obtain their energy from the sun are the primary producers and form the first of these trophic levels. The herbivorous animals which range from minute invertebrates to large mammals feed on these living plants, and are therefore described as being primary consumers; they occupy a second trophic level. Other animals prey upon the herbivores and form a third trophic level and, in turn, these carnivores are preyed upon by top or predatory carnivores to produce a fourth trophic level. The decomposers and detritus feeders form yet another trophic level, which accepts the residues from the other levels and turns them back into nutrients to be used by the primary producers, together with the sun's energy, to store new energy for the whole ecological system.

All ecosystems possess a characteristic trophic structure, and may therefore be studied by the investigation of that structure. In most systems, the only external source of energy is the sun, but, for many sub-systems, energy enters in the form of live or dead organisms, or in the form of decomposed organisms from another system. The energy is then used by organisms for synthesizing new compounds for growth and reproduction, and also to maintain the cells in their bodies, for movement, and to maintain body temperatures. While the energy for these processes can be made available through the breakdown of organic molecules in respiration, not all of the energy released in this way is utilized by the organisms, so that a proportion is lost and dissipated as heat. As a result, there is a constant flow of energy through the ecosystem from primary producers to carnivores and decomposers, and a constant loss of energy to the atmosphere as a by-product of respiration.

Change in ecological systems may therefore be assessed by an examination of the changes in the flows of energy. Such investigations formed an important part of the International Biological Programme (IBP) and many techniques were developed during that Programme for the measurement of biomass and the consequent changes of energy within different compartments of ecosystems. Without the kinds of models which are described in this Handbook, however, measurement of energy flow can never provide more than a static description of a particular ecosystem at a particular point of time. By the correct use of models, such descriptions may be developed to the point at which prediction can be made of the changes that will take place if the system is modified in some particular way.

It is not, however, solely the flow of energy to the tropic levels of an ecosystem which is of importance in the study of ecology. An almost equal interest is focused upon the flow of chemical elements, either as nutrients or as pollutants, through ecological systems. Primary producers take up mineral elements such as nitrogen and phosphorus, in the form of soluble mineral salts from surrounding soil or water. Herbivores and carnivores obtain nitrogen and phosphorus mainly as organic compounds in their food, though cattle and human beings may require supplements of raw minerals such as salt or copper. The dead organisms, together with their waste and excretory products, are broken down by decomposers (mainly bacteria and fungi) which release mineral nutrients in a form available for re-use by the primary producers. Similarly, there is a cycling of carbon released in the atmosphere as carbon dioxide as a result of respiration of plants, animals and decomposers. This carbon dioxide is taken up by green plants during photosynthesis, and the carbon passed on to animals when they eat the plants. There are similar cycles for nitrogen and oxygen and, indeed, for any chemical elements which we need to consider.

In recent times, it has become increasingly important to study the flow of pollutant elements and compounds through ecological systems. Many of these pollutants are stored and concentrated in animal tissues, reaching their highest levels in predatory animals, after successive concentrations at lower trophic levels. The mechanisms by which these substances pass through the trophic levels are therefore of particular significance, and it has become important to measure the changes taking place in such systems by the uptake and storage of elements and compounds which are otherwise foreign to the natural system.

Finally, the heritable characteristics of organisms are passed from parents to offspring by complex series of genetic events. Successive generations of organisms are subjected to varying pressures by their environment and by competing organisms; differentiated selection of genotypes thus induces change into the ecological system through the genetic make-up of the component organisms. It is, therefore, possible to describe dynamic change in ecosystems by documenting the genetic composition of organisms, and to predict future changes by modelling heritability and selection processes within defined populations. Similarly, it may frequently be possible to predict the effects of selection pressures upon populations of organisms through genetic mechanisms. Such models require particular information about the extent to which various characteristics of organisms are associated with genetic components, together with a knowledge of the fundamental laws of genetics.

In considering the dynamic change of ecosystems, we should not, of course, forget the important physical and chemical factors of the environment. Changes in these factors may require to be measured because of their impact upon organisms, or because they have themselves been altered by the organisms. Some of the more important factors may be related to the climate within which the organisms exist, either for a large area or for relatively small parts of the system, as in the soil.

Some dynamic changes involve regular cycles, as in the diurnal rhythms and in the seasons, but many of the changes are relatively unpredictable, as is the climate from year to year at a particular location. Other longer-term changes are associated with such factors as the melting of the ice in polar regions, sun spots, etc. Closely associated with climate are the factors of physiography which, particularly in Britain, have a strong influence upon the effects of climate through slope, aspect, and the drainage of soils. Similarly, the physical and chemical properties of soils are constantly changing, either through the effects of organisms themselves, or through deliberate attempts at management by man. The measurement and modelling of such changes become, therefore, an essential part of the modelling of dynamic change in ecosystems.

2.2 MEASUREMENT OF CHANGE IN ECOLOGICAL SYSTEMS

In the past, much of the measurement of change in ecological systems has been done through repeated survey. Such measurement requires an initial survey to provide a baseline against which change can be measured, followed by successive surveys to establish the direction and extent of change. There are, however, difficulties about the design of surveys to measure change, particularly when it is not clear what those changes may be. Thus, although a detailed survey may be made with the intention of providing the baseline for the monitoring or surveillance of change, and even though the precision of estimates made in the baseline survey may be high, the change itself may not be monitored with any great precision unless there is some clear hypothesis or hypotheses to direct the design of these surveys. Where, however, there are relevant hypotheses about the nature of the change, it may be possible to design repeated surveys which are capable of detecting change with reasonably high precision. Sampling with partial replacement, as developed for continuous forest inventory, is a particularly appropriate technique for use in such circumstances (Ware and Cunia, 1962; Cunia and Chevrou, 1969). A statistical checklist, highlighting some of the more important questions to be asked in the design of sample surveys, is given by Jeffers (1979).

Some authors make a distinction between monitoring and surveillance, depending on whether or not there is an attempt to correct the change taking place and bring it back to some stable state. No attempt will be made to maintain this distinction in this Handbook. Whether or not the ecological system is to be maintained in some preassigned state, it is necessary to determine the change which is taking place and to measure its extent. In essence, any investigation of dynamic change requires the definition and bounding of the problem to be framed as hypotheses which can be tested formally, even if that test can only be conducted after a chain of deductive reasoning from one or more hypotheses which are capable of direct verification (Jeffers, 1978a).

Three basic classes of hypotheses may be distinguished:

(i) Hypotheses of relevance identifying and defining the variables, species and sub-systems which are relevant to the problem.
(ii) Hypotheses of processes, linking the sub-systems within the problem, and defining the impacts imposed on the system.
(iii) Hypotheses of relationships, and of the formal representations of those relationships by linear, non-linear and interactive mathematical expressions.

These three classes of hypotheses may well be linked within a formal chain of deductive argument, leading to processes which can be summarized by a decision table enumerating all the hypotheses, and combinations of hypotheses, that must be specified in order to solve a particular problem. The decision table also specifies, for each combination of hypotheses, the decisions or actions that should be taken to ensure that the problem is correctly solved. Because decision tables provide a clear concise format for specifying a complex set of hypotheses and the various consequent courses of action, they are ideal for describing the conditions for interaction between component parts of a model. The extension of these techniques to the enumeration of the necessary combinations of hypotheses for particular courses of action where uncontrolled events may intervene, so that we are unable to control or predict with certainty, has been the main thrust of recent research into decision analysis (Raiffa, 1968).

The clearer definition of hypotheses about the nature and rate of change in ecological systems enables survey to be replaced by short-term and long-term experiments. The increased control over experimental areas provided by well-considered experimental design increases the precision with which change can be measured, and also offers the possibility of testing the interacting effects of various factors, including methods of management, conservation measures, and various forms of protective legislation. Indeed, by effective planning and design, it may be possible to provide a carefully assessed programme of environmental management linked to the detection of desirable or undesirable change in ecosystems. At the same time, unexpected change may be detected and included in subsequent measurements in the experimental areas. Regrettably, however, rather little emphasis has been given so far to the use of experiments as opposed to surveys taken as cross-sections in time, possibly because research workers are reluctant to commit their successors to maintaining the measurement and assessment of their experiments, or, alternatively, because research workers are reluctant to be committed to a programme of research by their predecessors.

It may well be that the academic and institutional organization of science, in developed and developing countries alike, precludes the rational design of investigations to determine and measure change in ecological systems over anything more than half a decade. A statistical checklist of the questions to be asked when designing experiments to detect change in ecological systems is given by Jeffers (1978b).

2.3 CONCEPTUAL MODELS OF DYNAMIC CHANGE

2.3.1 Vegetation

The species composition of all biological communities varies in time and space, the rate and character of the change being a function of the scale at which a community is examined, and of the external and internal factors which influence different populations in different ways. The interaction between organisms and environment that characterizes the point in time and space then becomes part of a larger system when arrays of sub-systems are combined into a biological community. It follows that communities and landscapes constitute a range of possibilities. These possibilities arise from combinations of the extent of differences between small sites and their patterns of diversity, the degree of biological influence on sites and on population regulation, the behaviour and longevity of dominant populations, the relative stability of populations, the roles of disturbance and succession, the kinds of succession and the extent to which species are replaced during successions.

Classical ecology has been mainly concerned with changes are reflected in vegetation, particularly in vegetation redevelopment following perturbation of some community. This emphasis arises because plants provide the energy base for all other biological activity, and because the vegetation provides the mechanical structure of the biological environment in which other organisms exist. The classical concepts of ecological succession involve two essential assumptions: first, that species replacement during succession occurs because populations tend to modify the environment, making conditions less favourable for their own persistence and leading to progressive substitution; and second, that a final and stable system, or climax, ultimately appears which is self-perpetuating, and is in balance with the physical and biological environment.

This view of ecological succession, developed over the centuries, and based largely on observed relationships between vegetation and environment, and on patterns of revegetation on agricultural and forest land, has been generally accepted by the scientific community, even though obvious exceptions to the form of succession exist. Egler (1954) was probably the first person to suggest formally that the classical model of succession may not apply in all situations, and he referred to this classical model as 'relay floristics' and suggested that, in many cases, the initial floristic composition following a perturbation may dominate the entire pattern of subsequent succession. He suggested that, unless species persisted throughout the perturbation, or were able to enter the perturbed site shortly afterwards, they would not subsequently be represented in the community that developed.

Evidence has gradually accumulated to suggest that the concept of the initial floristic composition may have wider applicability than was originally envisaged by Egler (Colinvaux, 1973; Drury and Nisbet, 1973; Horn, 1974). More recently, Connell and Slatyer (1977) have proposed a broader overall system of successional processes which incorporates the possibility of the pathways of relay floristics and initial floristic composition operating independently or in combination, and which, in addition, includes a pathway in which succession is truncated at a point short of the expected climax, a phenomenon which has been frequently observed (but seldom explicitly recognized) in successional theory.

In Figure 1, the various pathways are indicated, and the likely effects of perturbation, at different stages in each pathway, are also shown. The basic dichotomy between the classical 'relay floristics' concept (pathway 1) and the other models (pathways 2 and 3) is reflected in the immediate divergence of the pathways. In model 1, referred to by Connell and Slatyer as the 'facilitation' model, the classical replacement pattern occurs, each successive suite of species which occupies the site tending to make the environment less favourable for their own persistence and more favourable for their successors to invade and grow to maturity. In model 2, the 'tolerance' model, environmental modifications induced by earlier colonists may either increase or decrease the rates of recruitment and growth to maturity of later species. The latter appear later because they either arrived later, or, in present directly after the perturbation, had their germination inhibited and their growth suppressed.

In contrast to the 'facilitation' model, in model 3 — termed the 'inhibition' model by Connell and Slatyer — the early occupants, rather than facilitating the progressive occupancy by other species, inhibit the invasion of other species by pre-empting available space through physical occupancy, through physical competition, and the use of allelopathic substances, or through other effective means of inhibition. This inhibition has the effect of truncating the succession at a stage that would normally be regarded as being composed of

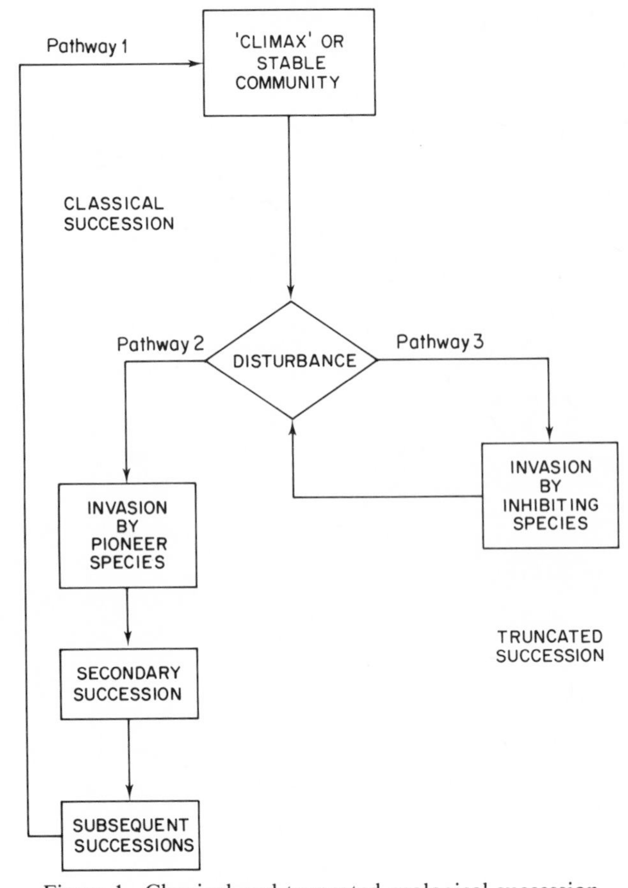

Figure 1. Classical and truncated ecological succession

non-climax species. Later succession species may only be able to enter the site when the inhibiting species are damaged or killed. If there is a subsequent perturbation, new succession may well follow a different pathway, avoiding a repetition of successional truncation.

In essence, the Connell and Slatyer concepts are based on the simple premise that the presence of a particular species in a community is dependent on the product of two probabilities. The first probability is that of a propagule being available at a site, itself a function of the ability to survive a pertubation or to reach the site by appropriate dispersal mechanisms. The second probability is that of the propagule being able to become established at the site and reach reproductive maturity, itself a function of the environmental requirements of the species, its adaptive ability and its reproductive strategy in relation to the prevailing environment.

2.3.2 Populations

Dynamic change in ecosystems may be measured directly by assessment of the numbers of plants and animals to be found in some convenient unit of space and time. The counting of numbers of plants per unit area at a particular instant of time is usually a straightforward, if tedious, task. Where the number of plants is large, sampling methods may be used to limit the numbers that have to be counted, and counts may also be limited to the larger and more conspicuous plants. Problems of identification, especially in the younger stages of most plants, will also frequently limit the extent to which a complete count of all species of plants occurring in a defined area is attempted.

The measurement of plants to determine the biomass of vegetation and the rates of growth of individual species and communities of plants is a more difficult task, and comprehensive procedures have been designed to ensure that effective standards are established for comparative measurements. The estimation of primary production of forests for the International Biological Programme is described by Newbould (1967) and the measurement of primary production of grassland is described by Milner and Hughes (1968). This Handbook will not attempt to repeat the information contained in texts on population assessment and measurement, but instead concentrates on the underlying mathematical models for the counts and measurements obtained.

Counts of animal populations and measurements of the growth rates of animals are even more difficult, if only because of the difficulties of catching the animals. It may also be necessary to identify, count, and measure animals at different stages of their development, particularly when the purpose of the investigation is to model the population dynamics of one or more species. Dempster (1975) emphasizes that a real understanding of the factors determining the abundance of animals can be obtained only by the intensive study of animal populations in the field, and describes the methods available for such studies.

The practical focus for models of population change will depend entirely on the hypotheses which are formulated for any particular investigation. For some studies, interest may be concentrated on only one species, but most studies are likely to be concerned with two or more species, as in the investigation of predator–prey relationships, or as in the elucidation of the complex interrelationships of competition between species for the critical factors of a habitat.

2.3.3 Communities

The study of ecosystems at the level of biomes has already been referred to in the discussion of the definition of ecological systems (Section 2.1). Change of one biome to another takes place only rarely, and probably only when there is a major climatic shift, or as a result of wholesale intervention by man. Within biomes, however, there are frequently smaller changes which are detectable because of the variations in the communities of plants and animals which are to be found at particular sites. Indeed, spatial variation or heterogeneity may provide broad indications of changes which will take place in time. Greatest interest is usually centred on those changes in communities of plants and animals which are related to measurable changes in the physical and chemical factors of the environment. In this way, a biogeographical interpretation of the dynamic and spatial variations may be achieved, leading to carefully structured hypotheses about the development of ecotones, climax states and seres. However, it is important to ensure that such hypotheses are capable of being tested, if the conceptual models of dynamic change are to be regarded as scientific as opposed to philosophical.

2.3.4 Management

By far the greatest influence on conceptual models of dynamic change in ecosystems is that of deliberate management by man to achieve particular aims. From earliest times, man has sought to maintain and improve his standard of living by hunting animals for food, clothing, and working materials such as bone, sinew, and horn. His pursuit and capture of animals for these purposes modified ecological systems already changing for other reasons, sometimes leading to the extinction of his prey. With increasing knowledge, man was able to domesticate some animals and to adjust his rates of cropping domestic and wild animals so as to sustain human populations in the modified ecosystems in which these animals lived. Nevertheless, unexpected changes still occurred, sometimes because of diseases and pests which changed the carefully contrived balance between the animals and the rest of the ecosystem, sometimes because of economic and social pressures which induced changes in demand and supply faster than the ecological system could itself respond.

Similarly, the harvesting and cropping of wild plants quickly led to methods of cultivation which have transformed whole ecosystems throughout the world. Ecologists sometimes pretend that these cultivated ecosystems are less interesting by comparison with natural or semi-natural ecosystems, but this view is an artefact of the perceptions induced by our urban-orientated education systems, leading to a higher valuation of the 'wild' and 'natural' than of the modified and cultivated. From the point of view of the complex interrelationships between plants and animals, and of both with their environment, changes in cultivated ecosystems are as ecologically interesting as those in natural or semi-natural ecosystems. Indeed, the study of such changes, in response to economic, social and technical initiatives, is at the heart of the applied ecology of agriculture and forestry.

While much of the management of cropped ecosystems is concerned with judicious disturbance, disease and pest control, and with regeneration, an especially important influence on dynamic changes in such ecosystems is that which is due to deliberate manipulation of the genetic composition of the crop plants or animals. Such manipulation may also be extended to weed species, and to pests and diseases. In natural and semi-natural ecosystems, genetic changes are usually relatively slow. In managed ecosystems, it is usually necessary to speed up these changes by programmes of plant or animal breeding. The new strains which result from these programmes then make new demands on the ecosystems for which they were created, inducing further modification and change.

2.4 ECOLOGICAL THEORY AND APPLICATIONS

2.4.1 Theoretical concepts

In general, therefore, our search for dynamic change in ecological systems will be directed by hypotheses which we will wish to formulate about the mechanisms which govern and induce such changes. Only rarely will we be able to devise methods of detecting changes which are totally unspecified; still less can we expect to develop models of dynamic changes which are capable of predicting changes that are completely unspecified. Nevertheless, our models may frequently predict changes which are counter-intuitive, or which are unexpected. It is these unexpected changes which provide a critical test for the hypotheses which underlie the models.

Much of the ecological theory which we will wish to use in the development of models of dynamic change will be related to trophic levels in ecosystems. In particular, the changes related to the flow of energy through the various trophic levels will provide a basic concept around which many of the models can be clustered and interrelated. Closely associated with the flows of energy will be the flows of nutrient and pollutant elements and compounds which may act as stimulants or inhibitants of growth, reproduction and distribution of organisms. The ultimate fate of many of these substances will help to determine the nature and extent of dynamic changes.

At a higher level, and therefore more visibly, dynamic changes in populations and communities of organisms, through responses to environmental changes (some of which have been induced by the organisms themselves), will often provide the basis for our models. Models of population dynamics may then be used to study the processes of ecological succession, although some problems of succession will be addressed directly by appropriate models. Finally, the changes related to evolution and genetic composition will necessarily be based on genetic models which consider the patterns and mechanisms of inheritance.

2.4.2 Application

The models of dynamic change described in this Handbook have a wide range of application. Obviously, they are especially relevant to many kinds of fundamental research in ecology. Indeed, many of us believe that little progress can be made in ecology unless mathematical models are used to describe the complex interrelationships between organisms, and between organisms and the physical and chemical factors of their environment.

There are, however, many fields of practical application for which these models are relevant. Agriculture and forestry, as two broad fields of applied ecology, have already been mentioned. Wildlife conservation, including both plants and animals, is another field of applied ecology where the modelling and prediction of dynamic change is essential if wise decisions are to be made about the conservation of specific organisms, habitats, or the management of nature reserves. The more direct effects of urban man in his pursuit of recreation and visual amenity also require to be examined by the kinds of models described in this Handbook.

However, perhaps the most obvious, and urgent, requirement for models of dynamic change in ecosystems is in the assessment of environmental impact. Many books have been written on this important topic in the last ten years (e.g. Munn, 1975), but the development of methods to assess the environmental impact of proposed human actions, such as the construction of large engineering works, land reform, and legislative policy, is totally dependent on our ability to model the dynamic change of the ecological systems upon which the actions will have an impact.

CHAPTER 3

Outline of Modelling Techniques and their Interrelationships

3.1 MODELS

In this Handbook, the word 'model' is taken to mean a formal expression of the essential elements of some problem in either physical or mathematical terms. In past scientific work, much of the emphasis in scientific explanation has been on the use of physical analogues of natural phenomena, and there is still a need for physical analogues in the study of biological and environmental processes. More generally, however, the models with which we are concerned here will be mathematical and essentially abstract.

We have explicitly excluded 'word models' or purely verbal representations of problems from further consideration in this Handbook. It is true, of course, that our first recognition of any problem is likely to be expressed in words, and there is much to be gained by seeking the most precise description that can be made of any problem with which we are concerned. It is surprising how often even four or five people closely concerned with a problem will disagree with each other's descriptions of the same ecological system, and disagreement on the particular elements of the systems which contribute, directly or indirectly, to a problem of practical concern is even more likely. For the larger groups of scientists, resource managers and administrators likely to be concerned with problems of dynamic change, the disagreement may be both striking and difficult to resolve. There is, therefore, all the more reason to spend some time in an attempt to find an agreed description, even if that description contains some passages, expressed as alternatives, for which no agreement can be reached.

Our reason for concentrating on the use of mathematical expressions as opposed to verbal expressions and physical analogues lies in the ability of mathematics to provide a symbolic logic which is capable of describing ideas, and particularly relationships, of very great complexity, while, at the same time, retaining a simplicity and parsimony of statement. The whole basis of mathematical notation rests on this economical description of relationships as a symbolic logic, and such description is 'formal' in the sense that it enables predictive statements to be derived from the relationships. Without the ability to predict the results of changes in one or more elements in the relationship, we could not regard our statements as belonging to science rather than to metaphysics or to literature.

The use of a mathematical notation in the modelling of complex systems is, therefore, an attempt to provide a representational symbolic logic which simplifies, but does not markedly distort, the underlying relationships. The use of symbolic logic, because it is essentially a simplification, necessarily gives an imperfect representation of reality, and must therefore be regarded as a caricature. Nevertheless, the various mathematical rules for manipulating the relationships enable predictions to be derived of changes which may be expected to occur with time in ecological systems as various component values of these systems are changed. These predictions, in turn, enable comparisons to be made between model systems and the real systems which they are intended to represent, and, in this way, to test the adequacy of the model against observations and data derived from the real world. This 'appeal to nature' is an essential part of the scientific method. Indeed, manipulation of the model system may itself suggest the experiments which are necessary to test the adequacy of the system.

The advantages of formal mathematical expressions as models are:

(i) they are precise and abstract;
(ii) they transfer information in a logical way; and
(iii) they act as an unambiguous medium of communication.

They are precise because they enable predictions to be made in such a way that these predictions can be checked against reality by experiment or by survey. They are abstract because the symbolic logic of mathematics extracts those elements, and only those elements, which are important to the deductive logic of the argument, thus eliminating all the extraneous meanings which may be attached to the words. Mathematical models transfer information from the whole body of knowledge of the behaviour of interrelationships to the particular problem being investigated, so that logically dependent arguments are derived without the necessity of repeating all the past research. Mathematical models also provide a valuable means of communication because of the unambiguity of the symbolic logic employed in mathematics, a medium of communication which is largely unaffected by the normal barriers of human language.

The disadvantage of mathematical models lie in the apparent complexity of the symbolic logic, at least to the non-mathematician. In part, this is a necessary complexity — if the problem under investigation is complex, it is likely, but not certain, that the mathematics needed to describe the problem will also be complex. There is also a certain opaqueness of mathematics, and the difficulty that many people have in translating from mathematical results to real life is not confined to non-mathematicians. It is, therefore, always important to ensure that the results of mathematical analysis are correctly interpreted and to translate solutions from mathematical formulae into everyday language.

Perhaps the greatest disadvantage of mathematical models, however, is the distortion that may be introduced into the solution of a problem by inflexible insistence on a particular model, even when it does not really fit the facts. The pursuit of mathematical models is sometimes intoxicating, to the extent that it is relatively easy for scientists to abandon the real world and to indulge in the use of mathematical languages for what are essentially abstract art forms. It is for this reason that we will insist, in this Handbook, that models are fully embedded in a broad framework of systems analysis.

3.2 SYSTEMS ANALYSIS

In the sense in which we shall use the term in this Handbook, 'systems analysis' is the orderly and logical organization of data and information into models, followed by the rigorous testing and exploration of these models necessary for their validation and improvement. Systems analysis provides a framework of thought designed to help decision-makers to choose a desirable course of action, or to predict the outcome of one or more courses of action that seem desirable to those who have to make decisions. In particularly favourable cases, the course of action that is indicated by the systems analysis will be the 'best' choice in some specified or defined way.

The aim of the broad framework of systems analysis outlined in the next chapter is to promote good decision-making in practical applications, and, in our case, in the ecology of dynamic systems. The framework is intended to focus and to force hard thinking about complex, and usually large, problems not amenable to solution by simpler methods of investigation, for example by direct experimentation or by survey. The special contribution of systems analysis lies: (i) in the identification of unanticipated factors and interactions that may subsequently prove to be important, (ii) in the forcing of modifications to experimental and survey procedures to include these factors and interactions, and (iii) in illuminating critical weaknesses in hypotheses and assumptions. Just as the scientific method, with its insistence on the testing of hypotheses through practical experiments and rigorous sampling procedures, provides the essential tools for advances in our knowledge of the phsyical world, systems analysis welds these tools into a flexible, but rigorous, exploration of complex phenomena.

The inherent complexity of ecological relationships, the characteristic variability of living organisms and the apparently unpredictable effects of deliberate modification of ecosystems by man necessitate an orderly and logical organization of ecologial research which goes beyond the sequential application of tests of simple hypotheses, although the 'appeal to nature' invoked by the experimental method necessarily remains at the heart of this organization. Applied systems analysis provides one possible format for this logical organization, a format in which the experimentation is embedded in a conscious attempt to model the system so that the complexity and the variability are retained in a form in which they are amenable to analysis. A further reason for the use of systems analysis in ecological research lies in the relatively long timescales which are required for that research. It is, therefore, necessary to ensure the greatest possible advance from each stage of the experimentation, and the models of systems analysis provide the necessary framework for such advances. Furthermore, the present state of ecology as a science, with its extremely scattered research effort over a wide field, urgently needs a unifying concept. Not only is there a marked incompatibility of the many existing theories, but the weakness of the assumptions behind these theories is largely unexplored, partly because the assumptions themselves have never been stated. Systems analysis can, therefore, be used as a filter of existing ideas. Theories which have been shown to be incompatible can be tested as alternative hypotheses, and the systems analysis itself will frequently suggest the critical experiments necessary to discriminate between these hypotheses.

3.3 FAMILIES OF MATHEMATICAL MODELS

While some of the general properties of mathematical models have been touched upon above, an experienced systems analyst can recognize broad 'families' of models, in much the same way that an experienced botanist is often able to identify the genus to which the plant belongs, even when he does not know the species. It may, therefore, be useful to review briefly some of the main families which can be recognized among mathematical models. The list is far from being exhaustive, and the categories are also not mutually exclusive. The classification is, however, probably sufficient to provide examples of the basic requirements of models applied to practical problems.

3.3.1 Functional relationships

Many ecological models are based on studies of systems dynamics which are themselves based on servo-mechanism theory, coupled with the use of high-speed digital computers to solve large numbers of equations in a short time. These equations are more or less complex mathematical descriptions of the operation of the system being simulated, and are in the form of expressions for levels of various types which change at rates controlled by decision functions. The level equations represent accumulations within the ecological system of such variables as weight, numbers of organisms and energy, and the rate equations govern the change of these levels with time. The decision functions represent the policies or rules, explicit or implicit, which are assumed to control the operation of the system.

The popularity of dynamic models of this kind arises from the flexibility of the models to describe systems operations, including non-linear responses of components to controlling variables and both positive and negative feedback. However, this flexibility has its disadvantages. It is, in any case, usually impossible to include equations for all the components of a system, as, even with modern computers, the simulation rapidly becomes too complex. It is, therefore, necessary to obtain an abstraction based on judgement and on assumptions as to which of the many components are those which control the operation of a system.

The application of systems dynamics in modelling involves three principal steps. First, it is necessary to identify the dynamic behaviour of the system of interest, and to formulate hypotheses about the interactions that create the behaviour. Second, a computer simulation model must be created in such a way that it replicates the essential elements of the behaviour and interactions identified as essential to the system. Third, when it has been confirmed that the behaviour of the model is sufficiently close to that of the real system, the model can be used to understand the cause of observed changes in the real system, and to suggest experiments to be carried out in the evaluation of potential courses of action.

Systems dynamic models have an intuitive appeal. The formulation of the models allows for considerable freedom from constraints and assumptions, and for the introduction of the non-linearity and feedback which are apparently characteristic of many ecological systems. The ecologist is thus able to mirror or mimic the behaviour of the system as he understands it, and gain some useful insight into the behaviour of the system as a result of changes in the parameters and driving variables. The power of computers to make large numbers of exact but small computations also enables the ecologist to replace the analytical techniques of integration by the less accurate, but easier, methods of difference equations. Furthermore, even when the values of parameters are unknown, relatively simple techniques exist to provide approximations for these parameters by sequential estimates, or even to use interpolations from tabulated functions. In particularly favourable cases, it may even be possible to test various hypotheses of parameters or functions.

The lack of a formal structure for such models and the freedom from constraints can, however, also be disadvantageous. For one thing, the behaviour of even quite simple dynamic models may be very difficult to predict, and it is easy to construct models whose behaviour, even within the practical limits of the input variables, is unstable or inconsistent with reality. Even more difficult, determination of the way in which such systems will behave frequently requires extensive and sophisticated experimentation. It is certainly always necessary to test the behaviour of such model in relation to the interaction of changes of two or more input variables, and seldom, if ever, sufficient to test the responses to changes in one variable at a time.

Further discussion of dynamic models is given in Chapter 5. To summarize, however, dynamic models may well be helpful in the early stages of the systems analysis of a complex and dynamic ecological problem, by concentrating attention on the basic relationships underlying the system, and by defining the variables and sub-systems that the investigator believes to be critical. In the later stages in the analysis, it will often be preferable to switch the main effort of analysis to one of the other families of models. It is precisely for this reason that systems analysis explicitly defines a phase of developing alternative solutions to the problem.

3.3.2 Matrix models

The family of dynamic models described in the previous section offers almost complete freedom to the investigator in the expression of those elements considered to be essential to the understanding of the underlying relationships between those variables and entities that have been identified in the description of the system. The models usually strive for 'reality' — a recognizable analogy between the mathematics and the physical, chemical, or biological processes — sometimes at the expense of mathematical elegance or convenience. The price paid for the 'reality' is frequently a necessity to multiply entities to account for relatively small variations in the behaviour of the system, or some difficulty in deriving unbiased or valid estimates of the model parameters.

Matrix models represent one family of models in which 'reality' is sacrificed to some extent in order to obtain the advantages of the particular mathematical properties of the formulation. The deductive logic of pure mathematics then enables the modeller to examine the consequences of his assumptions without the need for time-consuming 'experimentation' on the model, although computers may still be required for some of the computations.

Some of the most elegant of these matrix models are represented by the Leslie deterministic models predicting the future age structure of a population of animals from the present known age structure and assumed rates of survival and fecundity. Predator–prey systems, which sometimes show marked oscillations, can also be encompassed by matrix models, by a relatively simple exploitation of techniques for relating population size and survival. Seasonal and random environmental changes and the effects of time lags may similarly be incorporated, though the models necessarily become increasingly complex in formulation. Dynamic processes such as the cycling of nutrients and the flow of energy in ecosystems can also be modelled by the use of matrices, exploiting the natural compartmentation of the ecosystem into its species composition or into its trophic levels. Losses from the ecosystem are assumed to be the difference between the input and the sum of output from, and storage in, any one compartment.

Although matrix calculations are sometimes extensive, especially in matrix inversion and in the calculation of eigenvalues and eigenvectors, and will often require the use of computers, they are usually less difficult to program than those involved in dynamic models. Furthermore, the properties of the basic matrices of the models enable the modeller to exploit the deductive logic of pure mathematics. Matrix models, therefore, represent an important and neglected family of models in systems analysis. A particular type of matrix model which is especially useful in the modelling of dynamic change in ecological systems is that of the Markov models, in which the basic format is of a matrix of entries expressing and probabilities of the transition from one state to another at specified intervals. These models will be considered in further detail in Chapter 6.

3.3.3 Statistical models

The families of models so far considered have been mainly deterministic. That is to say, from a given starting point, the outcome of the model's response is necessarily the same and is predicted by the mathematical relationships incorporated in the model. Such models are necessarily mathematical analogues of physical processes in which there is a one-to-one correspondence between cause and effect. There is, however, a later development of mathematics which enables relationships to be expressed in terms of probabilities, and in which the outcome of a model's response is not certain. Models which incorporate probabilities are known as stochastic models, and such models are particularly valuable in simulating the variability and complexity of ecological systems.

Statistical models of this kind include the models of spatial patterns of organisms, the analysis of variance, multiple regression analysis, and the Markov models mentioned in the last section. Although, apart from the Markov models, we will not be primarily concerned with statistical models in this Handbook, some mention of their application will be made in the other chapters, and particularly in Chapter 7.

3.3.4 Multivariate models

A particular class of statistical models which has special relevance to the modelling of dynamic change in ecosystems is that of multivariate models, where it is necessary to consider changes between many variables, or variates. A variate is a quantity which may take any one of the values of a specified set with a specified relative frequency or probability. Such variates are sometimes also known as random variables and are to be regarded as defined not merely by a set of permissible values like any ordinary mathematical variable, but also by an associated frequency or probability function expressing how often those values appear in the situation under discussion. There are many situations in ecology and other applications of systems analysis where models have to capture the behaviour of more than one variate. These models are known collectively as 'multivariate' and are related to techniques know collectively as 'multivariate analysis' — an expression which is used rather loosely to denote the analysis of data which are multivariate, in the sense that each individual under investigation bears the values of p variates.

Broadly, these models may be divided into two main categories:

(i) those in which some variates are used to predict others; and
(ii) those in which all the variates are of the same kind, and no attempt is made to predict one set from the other.

For the latter, which may be broadly described as descriptive models, there is a further subdivision into those models in which all the inputs are quantitative, and those models in which at least some of the inputs are qualitative rather than quantitative. Predictive models, on the other hand, may first be subdivided according to the number of variates predicted, and then by whether or not all the predictors are quantitative. The use of such models in the investigation of dynamic change in ecosystems is described in Chapter 7.

3.3.5 Mathematical programming

The term 'mathematical'programming' describes a series of models whose aim is to find the maximum or minimum of some mathematical expression or function by setting values to certain variables which may be altered within defined limits. The underlying mathematics of these models was developed during the early application of mathematical techniques to practical problems in what has now come to be known as operational research. The simplest of these problems, in which the objective function and the constraints are all linear functions, can be solved relatively easily by standard methods. In essence, any inequalities in the constraints are first removed by introducing some additional 'slack' variables. Any feasible solution to the problem is then sought and, once such a solution has been found, iterative attempts are made to improve the solution, i.e. to move it closer to the defined optimum of the objective function by making small changes in the values of the variables. This iterative procedure continues until no further improvement can be made. Non-linearity in either the objective function or the constraints, or both, introduces disproportionate levels of difficulty in finding appropriate solutions. So, too, do problem formulations which impose limitations on the size of the lumps in which units of some particular resource can be allocated. There is, nevertheless, a well-developed theory of non-linear programming to cope with such problems.

As a further extension, some large optimization problems can be reformulated as a series of smaller problems, arranged in sequences of time or space, or both. A reformulation of this kind is frequently desirable in order to reduce the computational effort of finding a solution, although care has to be taken to ensure that the sum of the optimal solutions of the sub-problems approaches the optimum solution of the whole problem. Mathematical programming models are not widely used in the modelling of dynamic change in ecosystems and are not, therefore, further discussed in this Handbook.

3.3.6 Game theory models

Closely related to mathematical programming models are the models which are based on the theory of games. The simplest of these models is known as the two-person, zero-sum game. Such games are characterized by having two sets of interests represented, one of which may be nature or some external force, and by being 'closed' in the sense that what one player loses in the game the other must win. The theory can, however, be extended to many-person, non-zero-sum games. Such models represent an interesting, and so far little explored, alternative approach to the solution of strategic problems. The extension to the more complex non-zero-sum games and to many-person games, in which coalitions can be formed between the players, represents a field of research which deserves increased attention, particularly in ecological research related to the assessment of environmental impact and environmental planning. Because of a lack of practical examples of application of such models to the modelling of dynamic change in ecosystems, they will not be further discussed in this Handbook.

3.3.7 Catastrophe theory

The theory of catastrophes is an elegant development of mathematical topology applied to systems which have four basic properties, namely bimodality, discontinuity, hysteresis and divergence. Catastrophe theory models have attracted much interest and attention since they were first proposed in 1970. The models have considerable intellectual and visual appeal, but are not easy to apply in highly multivariate situations. There are also serious difficulties to be overcome in estimating the parameters of the model from ecological data. Again, although we are likely to see wider use of such models in the study of dynamic change in ecosystems, further discussion of such models will not be attempted in this Handbook.

3.4 INTERRELATIONSHIPS BETWEEN MODEL FAMILIES

It is clear that the list of families of models described above is far from being exhaustive, and that the categories are also not mutually exclusive. Thus,

Markov models belong both to the family of matrix models and to the family of statistical models. Furthermore, some basic statistical models are frequently essential to the development of dynamic models.

The distinction between deterministic and stochastic models is of particular importance. With a deterministic model, one will always arrive at the same predictions for given starting values, and for given values of the coefficients or parameters of the model. If, on the other hand, a stochastic model is used as a basis for simulation, the outcome of the simulation will not always be the same, even when the parameters and starting values are the same. The random elements in the model ensure variability, and the aim of such models is to mirror the variability found in living organisms and in ecological systems. As with experiments on the organisms themselves, it will usually be necessary to make repeated trials of the simulation in order to determine the ways in which the system will respond to various changes.

A further important distinction is related to the dimensionality of models, i.e. to the number of independent dimensions or variables that are included in the model system. Many variables will be inter-correlated and the actual number of independent dimensions will, therefore, be smaller than the total number of variables. One of the most valuable characteristics of the multivariate models described in Chapter 7 of this Handbook is that they help to determine the true dimensionality of the model system, and to select the most critical and useful variables upon which to base the simulation.

Again, while we have classified broad groups of models into families according to their mathematical characteristics, we could also classify models according to their purposes. Maynard-Smith (1974) makes a distinction between 'models' and 'simulations'. He regards a mathematical description with a practical purpose, which includes as much relevant detail as possible, as a 'simulation', and restricts the use of the word 'model' to descriptions of general ideas which include a minimum of detail. This is not a distinction which will be maintained in this Handbook.

Many of the models that we may wish to construct of dynamic change in ecological systems may be regarded as *descriptive models*. Their aim is to provide as good a description as possible of the underlying processes or relationships on which dynamic change depends. Such models may help to weld together widely disparate theories about ecological systems, or, alternatively, they may help to show the incompatibility of commonly held beliefs or theories. Once formulated, the descriptive model will be used to obtain a further understanding of the way in which a single organism, a community of organisms, or a whole ecosystem will respond to various changes, natural or induced.

Alternatively, a model may be described primarily for the purpose of prediction. Such *predictive models* may pay relatively little attention to the physics, chemistry or biology of the underlying processes, but will be regarded as efficient only if they enable predictions of future states of the system to be made with a known degree of accuracy. Predictive models may be derived as the operational versions of descriptive models, the improved knowledge of the underlying processes being used to refine the predictive capability of the mathematical expressions. It may, nevertheless, be possible to develop predictive models directly, sometimes with drastic simplification of the mathematical assumptions about ecological processes. Efficient descriptions and efficient predictions are not necessarily closely related.

A third class of models may be distinguished as *decision models*. the aim of such models is neither to provide a description of the ecological system nor to predict the future state of such a system, but, instead, to guide practical decisions about the management or treatment of the system. In a sense, of course, both descriptive and predictive models can be used to guide practical decisions. Decision models are, however, specially formulated so as to provide such guidance more directly, and, in addition, by showing the consequences of particular choices about the management of an ecological system, they help to indicate a management strategy which is 'best' in some predefined way. Decision models are not usually derived from descriptive or predictive models, but are developed from families of models which have distinctive mathematical properties, like mathematical programming.

3.5 CHOICE OF MODELS FOR DYNAMIC CHANGE OF ECOSYSTEMS

As this Handbook cannot be a comprehensive guide to the whole set of model families, a limited number of models have been chosen as being of greatest value in the modelling of dynamic change in ecosystems. Three main classes of models are described in detail, namely dynamic or functional models, Markov models, and multivariate models. For all three classes, there is now sufficient experience of their application to the modelling of dynamic change to enable an assessment to be made of their usefulness, and of the difficulties of constructing models from the kinds of data likely to be available. Future revisions of this Handbook will almost certainly include other classes of models as expertise is developed in their use and construction.

CHAPTER 4

Systems Analysis

4.1 INTRODUCTION

In the previous chapter, it was emphasized that mathematical models should be incorporated into a broad framework of systems analysis. In that chapter, models were defined as formal expressions of the essential elements of a problem in either physical or mathematical terms. Similarly, systems analysis was defined as the orderly and logical organization of data and information into models, followed by the rigorous testing and exploration of these models necessary for their validation and improvement. It is on this orderly and logical organization, and the subsequent validation and testing of models, that we shall concentrate in this chapter. We will also review the role of system analysis in:

(i) integrating research on complex problems;
(ii) providing a link between research and the application of research results.

4.2 MAB CONCEPT OF SYSTEMS ANALYSIS

The earliest detailed consideration of systems analysis and modelling in the Man and the Biosphere Programme of UNESCO was at a meeting of an Expert Panel on Systems Analysis and Modelling Approaches, which took place in Paris in April 1972 (UNESCO, 1972).

After considerable discussion by the experts on this Panel, it was concluded that integration and co-ordination of scientific activity are needed to bring about greater coherence of ecological research, and that systems analysis could help to bring about this coherence, and could increase the significance of the research for the practical management of resources.

Systems analysis was not formally defined by the Expert Panel, but the members' interpretation of systems analysis may be derived from the following considerations. First, there was a need to develop a predictive understanding of the functioning of the complex natural systems upon which man depends. When faced with complex and highly interactive systems, human judgement and intuition may lead to wrong decisions, sometimes with results that cannot be reversed. Numerous examples of these wrong decisions exist in the history of man's management of his natural resources, and, if anything, the number and the gravity of these errors have increased in the recent past. Mere increase of knowledge, without predictive understanding of the functioning of complex systems, is not sufficient for the management of such systems.

There has been extensive study of the behaviour of complex interacting systems in such fields as engineering, physiology and economics. Resulting from this study has been the development of methods for understanding the dynamics of systems, and the impact of stresses upon them. Such methods can be adapted to systems, ecology, with the assumptions that the state of an ecosystem at any particular time can be expressed quantitatively, and that changes in the system can be described by mathematical expressions. Understanding of any system depends on translation of its variables and properties into a generalized form so that is becomes an abstract, or model. Such models are essentially simplifications of the system, but are nevertheless more comprehensive and more precise than the mental models of the field scientist or the resource manager.

Figure 2 shows how mathematical models of different types can help bridge the gap between ecological theory, management experience, experiments, and between strategic and tactical management prescriptions. The analysis incorporates data and information from a wide array of studies into a single interrelated system. Analytical procedures are used to select the combination of tactics required to give maximum output from a system consistent with a variety of constraints on management alternatives.

It is possible to recognize some common stages in the use of systems analysis for the solution of problems concerned with dynamic change in ecosystems.

4.2.1 Setting of objectives and preliminary synthesis

Such studies necessarily begin with the establishment of objectives and the creation of an initial synthesis of existing information. The objectives should specify: (i) the range of subject matter; (ii) the types of manipulation, modification or disturbance to be included in the prediction; and (iii) the variables which it is intended to measure and to predict. The initial synthesis requires the assembly of relevant existing knowledge from publications, unpublished reports and field data, and discussions with experienced research and management personnel. Assumptions will necessarily have to be made as to the relevance of particular kinds of knowledge, particularly when such knowledge exists at the periphery of the principal problem. For this reason, it is frequently useful to re-examine the objectives after the preliminary synthesis has been completed, as the act of assembling data will have helped to clarify ideas in the minds of the research scientist. Whenever possible, the existing knowledge should be incorporated into predictive models as a basis for the design of the next phase of the investigation, which is concerned with direct experimentation.

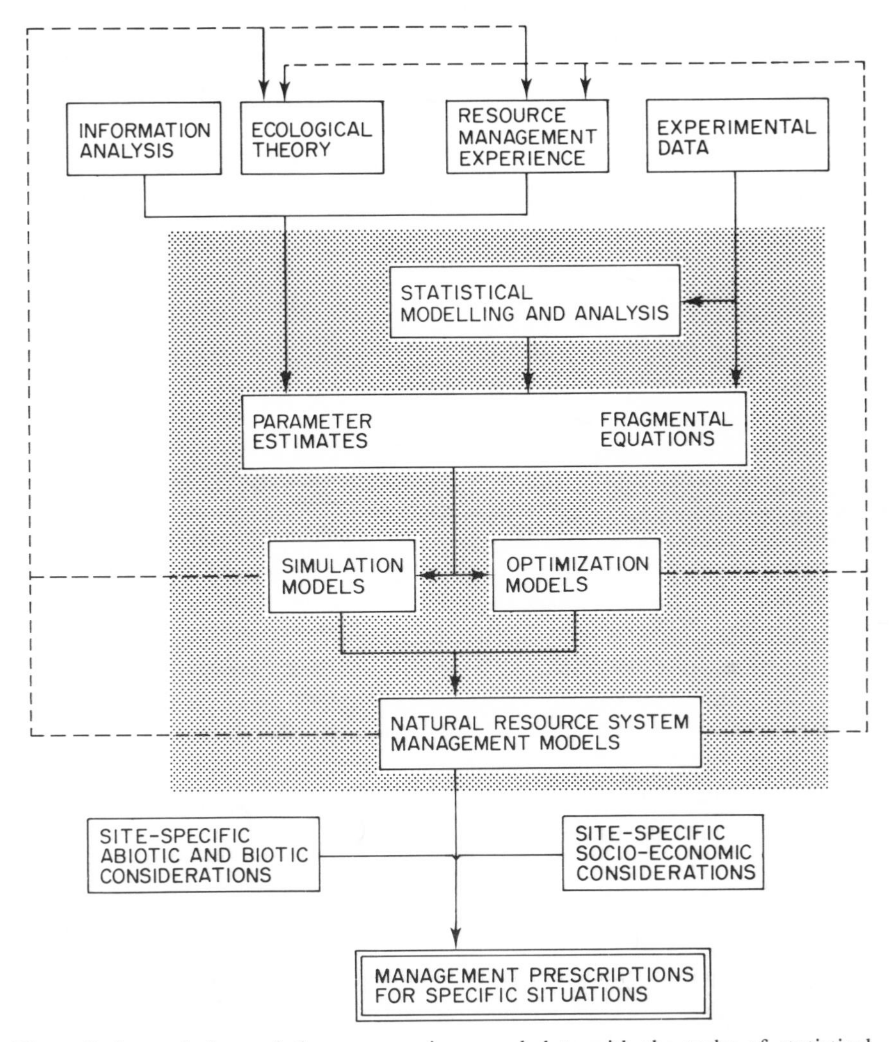

Figure 2. Interrelations of theory, experience and data with the tasks of statistical, simulation and optimization modelling towards improved recommendations for natural resource management. Solid lines represent direct flows of information, dashed lines represent feedbacks. The central (shaded) portion of the figure includes many of the kinds of models used in a systems analysis and operational research approach. The elements of the more conventional approach to research management decision-making lie outside the dotted portion

4.2.2 Experimentation

Once the objectives of the project and the preliminary synthesis have been completed, it will usually be necessary to proceed to direct experiments in the field and in the laboratory, as well as on any preliminary models which may have been developed. The type of the experimentation depends principally on the level of sophistication of the project. Validation data should therefore be gathered to test the output of, or the functions in, the predictive models, so that these models are themselves helping to define the kinds of experiments which need to be performed.

Adjustment or revision of models is greatly accelerated by formulating procedures for the rapid transmission of information on the structure and operation of the models which have been developed. Similarly, experiments conducted on models leads to new ideas on the management of the system being examined. It is therefore necessary to ensure a three-way communication and feedback between research scientists, modellers and managers or planners. Ideally, this communication should be initiated earlier during the phases of synthesis, planning and setting of objectives, so that all three groups of personnel have an opportunity to participate in the development of the research from the beginning. To aid to this communication, it is also a useful procedure to develop simulation and optimization models of natural resource systems for use as models in management games.

4.2.3 Management

The communication and dialogue between scientists, managers and planners during the examination of the output from models and experiments should result in the planning of pilot-scale management studies. After appropriate testing at the pilot scale, larger-scale management schemes can then usually be undertaken, again with consultation and feedback between the scientific groups developing the synthesis, experiments and models.

4.2.4 Evaluation

As the research proceeds, it is possible to evaluate the effects of the management changes proposed on the structure, function and stability of the system. In these terms, evaluation is a continuous process, starting with the setting of objectives and priorities, but, at an appropriate stage, it may be desirable to summarize the results of all the ideas developed in the earlier stages of the project. Such an evaluation may lead to a return to one of the earlier phases in order to correct for faulty assumptions or design failures in the model.

4.2.5 Final synthesis

Ultimately, it is necessary that the information derived from the research should be summarized, evaluated and published. However, no model is ever likely to be regarded as 'final' so that the models which are derived from one research project are very likely to form the preliminary synthesis for further research.

With the development of computers and computer languages, new approaches and methodologies have been designed to handle complex biological systems, so that it is now possible to suggest new research and policy strategies for situations with a large number of interacting components. Nevertheless, the Expert Panel suggested that inadequate integration might be expected to occur in the systems analysis of ecological research at the following junctions:

(i) between data experimentation and model development;
(ii) between simulation model operations, the overall systems analysis and the implementation of models in field testing;
(iii) between the examination of predictions from systems analysis and the implementation of management procedures;
(iv) between the testing of management techniques and the development of new hypotheses;
(v) between the implementation of results from pilot studies and the development of new hypotheses.

The Expert Panel also presented the following conclusions.

(i) Data quality and the understanding of causal pathways in ecology are generally unreal, and this lack of reality is likely to hinder systems analysis of natural systems.
(ii) Systems analysts and data collectors can often develop a mutually beneficial relationship from which a decision-maker himself eventually derives the maximum benefits.
(iii) Systems training is valuable for stressing a broad interdisciplinary, problem-orientated philosophy of research, and such a philosophy is urgently needed for the kinds of ecolgical research which is necessary in the MAB programme.
(iv) Systems models can be improved only by building them, and by striving to correct the weaknesses of the early versions of these models. An adequate model does not spring fully fashioned from an initial research project.
(v) The scientists contributing to systems analysis must be broadly interdisciplinary, and the contributions of these many disciplines are essential for the development of the models of dynamic change in ecosystems.

(vi) Systems models often demand a large quantity of high-quality data, and they can therefore be very expensive. Nevertheless, systems analysis provides a powerful tool for prediction and planning, a set of procedures for the formal application of logical processes, and a means of communicating with research scientists and managers.

4.3 LATER DEVELOPMENT OF SYSTEMS ANALYSIS AND MODELLING

Since the meeting of the Expert Panel in 1972, there has been considerable development of systems analysis and modelling within the scientific community. In particular, the studies of the International Institute for Applied Systems Analysis in Laxenburg, Vienna, have further helped to define the role and broad functions of systems analysis. From this development, the role of systems analysis has come to be seen as an explicit formal enquiry, designed to assist in the making of decisions for the forming of policy. The analysis determines a preferred action or policy of identifying and examining the alternatives, and by a comparison of alternatives in terms of their consequences. Explicit formal enquiry usually involves the use of mathematical models, but such models are not strictly necessary.

From this definition, the broad functions of systems analysis may be defined. First, systems analysis provides an objective basis for assessing and assimilating available information about the system. Second, it directs research into areas for which, relative to the understanding of the whole system, present knowledge is uncertain. Third, it provides a means of assessing and applying the results of this research. Fourth, it assists in the management, control or development of the ecological system.

Figure 3 shows, in diagrammatic form, the seven steps which have been identified in the application of systems analysis to practical research. The process begins with the *recognition* of the existence of a problem, or of a constellation of interconnected problems. Such problems must be amenable to analysis and sufficiently important for detailed investigation, so that recognition is a critical step which may determine the success or failure of the subsequent research. The second step is the *definition and bounding of the extent of the problem*. This definition and bounding seeks a simplification of the problem in order to make it capable of analytical solution, while preserving all the elements which are of sufficient interest for practical research. Again, this is a critical stage in any analysis, and specialist experience is often valuable in helping to determine the relative importance of the inclusion or exclusion of elements of the problem. Such experience also facilitates the balancing of the relevance of elements of the analytical solution against complications which will make the solution unmanageable. The third step is the *identification of the hierarchy of goals and objectives*, in which the major objectives set in the

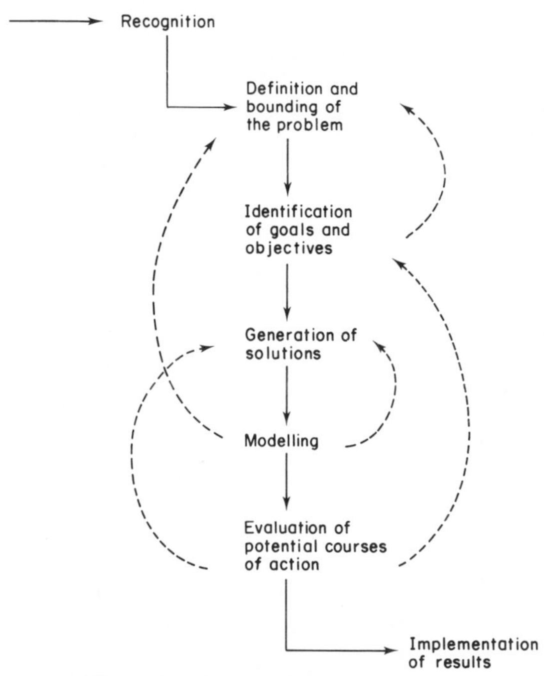

Figure 3. The steps in systems analysis

earlier stages are progressively subdivided into a series of minor objectives. Each of these minor objectives will frequently represent a 'milestone' in the conduct of the investigation. For this reason, it is necessary to ensure that the determination of the priorities is relative to the amount of effort required to meet the objectives. There is little point in placing a great deal of emphasis and effort on an objective which is only of minor importance. Conversely, most of the effort in the investigation should be directed towards the major objectives.

When these earlier stages have been completed, it is possible to proceed to the *generation of the solutions* to the problem. The aim is to generate a series of possible solutions, ideally selected as members of broad families of models, as described in the outline of modelling in Chapter 3. In generating solutions in this way, it is desirable to seek analytical models of the greatest possible generality. Such models will often make the best use of previous work on similar problems and, at the same time, will attempt to keep the underlying mathematics as simple as possible. It is seldom possible to predict the modelling strategy which is most likely to be successful in finding a solution to a particular practical problem, and as many models as possible should be developed in the early stages of the systems analysis. Only when some of the preliminary models have been tested fairly extensively will it be possible to decide on the best strategies for solving the problem.

Modelling of the complex, dynamic interrelationships of the various facets of a problem may be attempted for many of the alternative strategies. A full awareness of the inherent uncertainities in the various processes to be modelled, or of feedback mechanisms which may complicate both understanding and practicability of the model, can often be gained by practical experience. An experienced analyst may be able to contribute much to the shortening of this phase of systems analysis. There may, too, be a complex series of rules which will be used to reach a decision about appropriate courses of action. These rules must also be incorporated into the model.

Evaluation of potential courses of action for solving a practical problem will begin by investigating the sensitivity of the results of the modelling to assumptions made by the model, a process which is known to analysts as sensitivity analysis. In this analysis, previously unexpected weaknesses in the assumptions, and in the model formulation, may be revealed. Discovery of an important flaw in a major assumption may lead to a return to the modelling phase, but, sometimes, relatively simple modifications of the original model may be sufficient. Similarly, investigations of the sensitivity of the model to facets of the problem which were excluded from the formal analysis when the problem was defined and bounded will frequently play an important part in this evaluation. The problems of sensitivity analysis are discussed further below. Finally, however, systems analysis is incomplete unless the whole analysis is moved to the *implementation of results*. Having investigated and modelled the practical process, implementation may itself demonstrate that various phases of the analysis were incomplete or need to be revised.

Because systems analysis of a framework of thought rather than a defined prescription, the list of steps above needs to be understood in a qualified sense. Not all of the steps need to be included in every example of systems analysis. Similarly, the order in which the phases are undertaken may be varied, or it may be necessary to work through them in various patterns. For example, the importance of excluded factors may be reassessed repeatedly, necessitating several cycles of the modelling and evaluating phases. The relevance of the objective structure of the analysis may have to be examined periodically, sometimes requiring a return to one of the earlier phases, even after a considerable amount of work has been done on some of the middle and later phases. The most useful models will mimic reality with sufficient precision to serve a broad spectrum of decision and decision-makers. The decision phase may, therefore, be diffuse and broad, and follow the completion of the formal scientific analysis. Some of the possible points of return to earlier phases are shown in Figure 3.

4.4 SENSITIVITY ANALYSIS

As described above, sensitivity analysis involves the investigation of the effects of changes in the input variables and parameters on the results obtained from the models. Large or small variations resulting from such changes in performance of the model help to determine the 'sensitivity' of the particular variables and parameters, and so provide a basis of comparison for the importance of such elements. Ideally, sensitivity analysis begins during the modelling phase and continues to the end of the research and the implementation of results. Those parameters to which the model behaviour is especially sensitive can often be made the subject of close scrutiny and subsequent modification, with further experimental work or data analysis to ensure that those processes are more precisely modelled. Indeed, sensitivity analysis may itself aid decisions about allocations of resources to various parts of the research programme.

Sensitivity analysis may be expected to concentrate on four major aspects of model performance, and, specifically, to investigate uncertainty which has been introduced during the modelling phase. First, such an analysis will concentrate on the relative uncertainty of parameter values used in the model. Few of the ecological parameters necessary for the modelling of dynamic systems are known exactly. They frequently have to be estimated from the results of earlier research, or as a result of special surveys and experiments. At best, there will be sample determinations which are assumed to be representative of some population parameter. Second, parameter values may be expected to be subject to experimental variation, partly reflecting the inherent variability of the biological material modelling in the dynamic system, and partly dependent on unavoidable differences in the ways in which successive trials and experiments are carried out. Experimental variation of this kind may have a significant influence upon the range of values within which the true values of the parameters lie, especially where the values are themselves not independent from one variable or parameter to another. Third, the importance of interactions between parameter values and the effects of significant variables within the dynamic system can seldom by overemphasized. Few, if any, models of dynamic systems can be reduced to such simplicity that the behaviour of the model systems is adequately defined by models with completely independent variables. The investigation of interactions in systems analysis makes it essential to perform such analyses by varying more than one variable at a time in the model system. Fourth, it is frequently necessary to extend the range of input state variables to ensure that the model behaves consistently for the full range of values for which predictions are to be made. Most model systems represent an extrapolation from the range of conditions and processes for which information is available experimentally, even when the scientists constructing the model have been careful to stress the dangers of extrapolation. It is important, therefore, to know the range of variables for which the model behaves in a reasonable manner, so as to ensure that those subsequently using the model do not extrapolate beyond these points.

As will be apparent, there are important implications of sensitivity analysis in the whole process of modelling. For example, sensitive regions of the model may indicate sensitive regions of the real system, and such regions may need to be stressed in terms of closer control in the implementation of results from systems analysis, and in the preparation of guidelines of management. Similarly, the process of sensitivity analysis may indicate the need for close validation of specific sub-systems, relationships or individual parameters. In particular, the analysis may reveal overemphasis of particular processes where these processes are themselves of relatively little importance in the overall system. In this way, it may be possible to achieve modification of the model required to give a more balanced representation of the real system. Again, as has previously been suggested, the analysis may indicate the sensitive areas or elements which have been isolated by sensitivity analysis and which need to be better understood and properly represented. Such a process again helps to establish research priorities.

4.5 VERIFICATION AND VALIDATION

Although many texts do not make a distinction between verification and validation, it is often helpful to do so, although the usage of these terms is certainly not consistent. Verification may be regarded as the process of testing whether the general behaviour of a model is a 'reasonable' representation of that part of the real-life system which is being investigated, and whether the mechanisms incorporated in the model coincide with the known mechanisms of the system. Verification is, therefore, a largely subjective assessment of the success of the modelling, rather than an explicit test of the hypothesis underlying the model. Some verification will inevitably have been going on during the hectic phase of mathematical activity, as the 'reasonableness' of the results will be one of the criteria by which the modeller will have judged the success or failure of his efforts. Nevertheless, what is reasonable in small parts of the model may be less so when the parts are put together into a composite of the individual parts. Interactions between responses and impacts may need to be explored sequentially and factorially to ensure that the full range of possible conditions have been covered, and that, within the limits bounding and defining the problem and the ecosystem, and model behaves, for the defining purposes, in much the same way as the real system. We must, of course, be careful that we do not reject a model simply because it behaves in a counter-intuitive fashion. There are plenty of examples of solutions which are contrary to what is usually regarded as common sense. No model should therefore be rejected simply because the results are unexpected. Where, however, the model behaves in a completely different fashion from the real system which is being investigated, some explanation has to be sought, at the very least, for the inconsistency. This is the role of verification.

Validation, in contrast, is the quantitative expression of the extent to which the output of the model agrees with the behaviour of the real-life system, and is an explicit and objective test of the basic hypotheses, made by means of a delineation of test procedures, primarily statistical, which are applicable to the determination of the adequacy of the model. In most ecological applications of systems analysis, this process of validation has hardly been attempted, mainly because of inadequate definition and bounding of the initial problem. Typically, validation, where it is attempted at all, is approached in a direct and obvious manner, mainly by observing the behaviour of the model systems under a set of controlled or measureable loading and other conditions and then comparing the observation with corresponding predictions of the simulator. When the observations and the predictions agree within the required limits for all conditions treated, the simulation is considered to be validated.

The procedure of validation has several recognized difficulties, not the least of which is the uncertainty associated with the drawing of general conclusions from a finite, and typically small, number of experiments. This uncertainty is of particular concern in the validation of systems analysis models where one may be attempting to predict effects which are of the same order of magnitude as the random fluctuation or 'noise' inherent in real system measurements. In such cases, it is advantageous to use techniques for statistical design and analysis of experiments, both to reduce the number of experiments needed for a given level of confidence and to indicate the statistical significance of measured and simulated effects. Fortunately, effective techniques for experimental design have been developed during the last fifty years, and techniques which were originally intended for use in experiments on real-life systems are now proving valuable in testing the behaviour of simulations of those systems. A full account of these techniques is given by Schatzoff and Tillman (1975), Kleijnen (1975), and Dent and Blackie (1979).

4.6 THE VALUE OF SYSTEMS ANALYSIS

Systems analysis does not merely help to define priorities. The modelling process itself indicates the exact form in which data can be most readily used, so that simulation can assist in directing the tactics, as well as the strategy, of research planning. Similarly, while the results from the experiments may be published and disseminated by traditional forms of publication, they may also be assimilated and assessed within the models of systems analysis. As these models may eventually by incorporated into higher models as sub-models, they may be expected to have two major tasks, namely that of guiding the establishment of research priorities, and as a medium for the accumulation of research findings.

The introduction of systems analysis and systems modelling in ecological research also has a major effect upon the direction of research. Ecological research is costly, and the results are sometimes uncertain. Management of such research therefore requires objective guidance in establishing programmes for maximum effectiveness, and such guidance may be provided by systems analysis. If systems analysis is directed towards the ecological system, the objective of the better understanding of the real-life system so that is can be more effectively monitored and controlled helps to provide a linkage between research and the applied system. Such a linkage is illustrated in Figure 4.

Following Rowen (1976), good systems analysis has the following characteristics.

(i) The analysis uses methods which fit the character of the problem and the nature of the available data, while treating all data sceptically.
(ii) Systems analysis defines, explores and reformulates objectives, while recognizing that there may be several objectives capable of being arranged in a hierarchy.
(iii) Good systems analysis uses criteria sensitively and with caution, giving weight to qualitative as well as quantitative factors.
(iv) Effective analysis emphasizes design and creation of alternative solutions and options, and avoids concentration on too narrow a set of options.
(v) Modelling within systems analysis should handle uncertainty and stochastic variables explicitly.
(vi) The approach to the problem should indicate that the analyst understands the essential nature of the practical problems.
(vii) It is important to use simple models to simulate the essential aspects of the problem, and to avoid large and complex models that attempt to mimic reality while concealing the basic structure of the problem and the uncertainties of the estimation of model parameters.
(viii) The results should display honesty in the labelling of assumptions, values, uncertainties, hypotheses and conjectures.
(ix) The whole process of systems analysis should demonstrate understanding. The task is not merely to indicate the 'best' solution, but also to develop a range of alternatives.
(x) The analysis and its results should also show that an effort has been made to understand the practical problems and constraints of management and administration, especially if the analysis suggests a radical reformulation of the problem.
(xi) The solutions should take into account the organizational factors that affect the alternatives generated and influence the decisions.
(xii) The whole procedure of systems analysis should exhibit awareness of partial analysis, and the limits of analysis generally.

These additional precepts have been suggested as a result of experience in IIASA.

(xiii) Good systems analysis makes as certain as possible that the suggested alternatives are feasible.

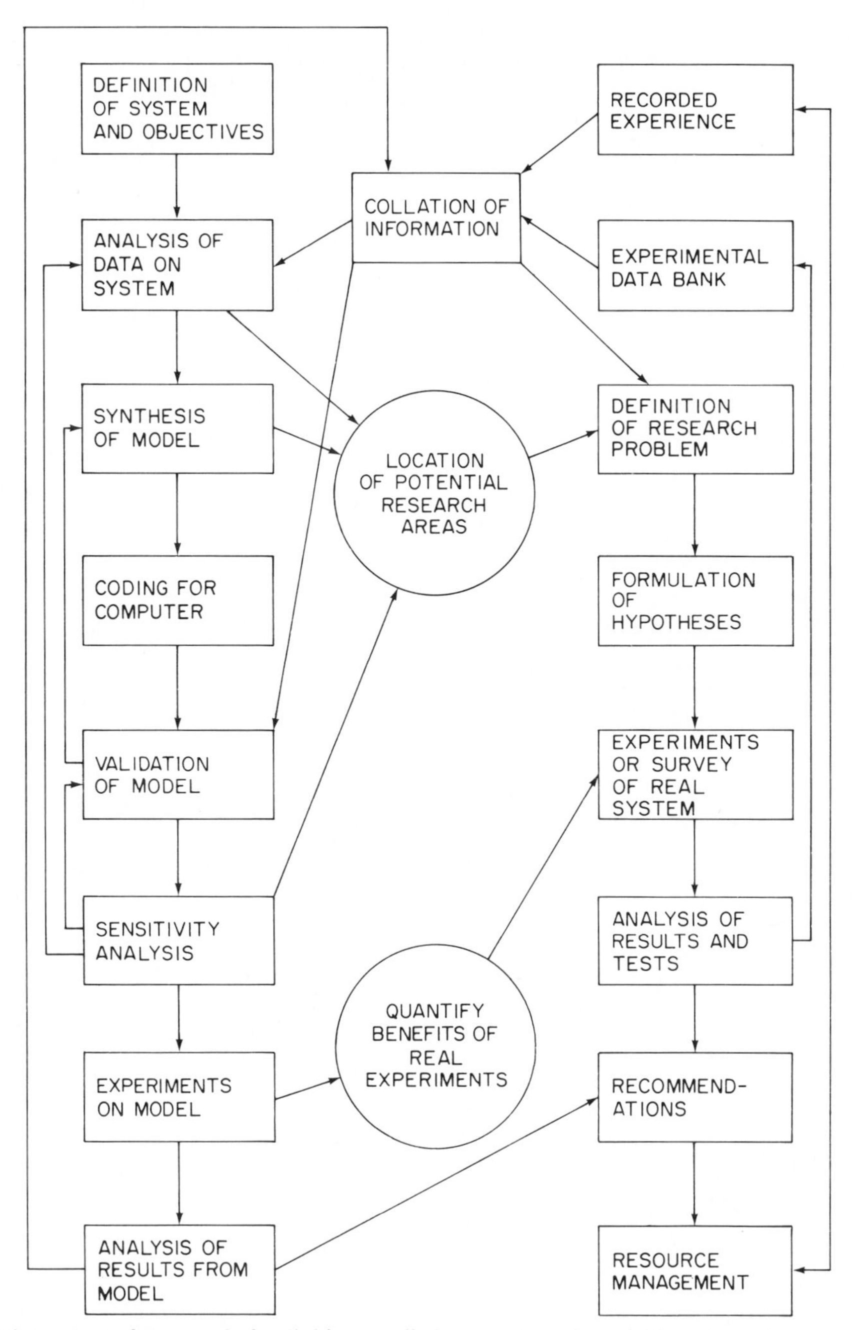

Figure 4. A framework for linking applied research and modelling through systems analysis (after Dent and Blackie, 1979)

(xiv) The analysis should consider the difficulties of the implementation of solutions and the costs of achieving them.

(xv) The analysis should recognize that an approximate solution before any decision has to be made is better than an exact solution long after the decision has been made.

CHAPTER 5

Dynamic Models and Sub-models

5.1 THE ORIGINS AND DEVELOPMENT OF DYNAMIC MODELS

Today's dynamic models and sub-models of ecosystems have their origins in the classical analytical models of physics, chemistry and biology. These analytical models depend on mathematical procedures for finding exact solutions to differential and other equations. As the behaviour of many kinds of mathematical equations is well known, such analytical solutions provide considerable scope for the investigation of the response of mathematical expressions to changes in basic parameters or variables. They require little more than the use of pencil and paper and a knowledge of mathematics, although that knowledge may need to be considerable for anything but the simplest of problems. Analytical solutions to such problems have the enormous intellectual appeal which characterizes the aesethetics of mathematics, as well as the predictive power of scientific hypotheses. Indeed, Maynard-Smith (1974) makes a clear distinction between what he calls models, i.e. mathematical descriptions of ecological systems which are essentially theoretical, and simulations, which have a practical purpose and whose utility lies mainly in the description of particular cases.

Undoubtedly, analytical models have been, and remain, useful in some fields of ecology, especially in population biology and population dynamics, as well as in the development of the theoretical models which are the basis for many developments in ecological theory. A simple example of the use of analytical models is provided by the Volterra equations describing the interactions between a prey species and its predator. These equations were described by Volterra (1926), and Maynard-Smith (1974) gives the following analysis.

> If we assume that the density of the prey species is represented by the variable x, and the density of the predator species by the variable y, then one model for the interactions between predator and prey is given by the equations:
>
> $$\dot{x} = ax - bx^2 - cxy$$
> $$\dot{y} = -ey + c'xy$$

This model makes a number of important assumptions, which may be summarized as follows.

(i) The density of a species (the number of individuals per unit area) can be adequately represented by a single variable.
(ii) Changes in density can be described adequately by deterministic equations.
(iii) The effects of interactions within and between species are instantaneous.
(iv) In the absence of predation, the prey species would follow the logistic function, with an intrinsic rate of increase and a carrying capacity of a/b.
(v) The rate at which prey are eaten as proportional to the product of the densities of predator and prey.

How does the system described by these equations behave? At any given time, t, the state of the system is described by the values of x and y, so that there is a corresponding point describing the system in the x and y phase plane. To each point we can also attach an arrow to indicate the direction in which a system at that point will move, and, if these arrows are joined up, they form trajectories describing how the system behaves. Formal analysis of the system in this way, derived from knowledge of the properties of differential equations, demonstrates that the system converges to a steady state with prey and predator both present only when the carrying capacity a/b is high enough to support the predator. Modification of the original equations enables various other influences on the relationship between prey and predator to be investigated, for example the effects of cover for the prey, the effect of a predator with a constant food intake, etc.

See Manynard-Smith, 1974, for a more complete account.

Despite their aesthetic attractions, analytical models are useful only under certain rather restricted conditions. First, the equations describing the ecological processes must be linear if more than a small number of these equations are to be solved simultaneously. Second, the solution of even a small number of non-linear equations may become difficult or even impossible under particular cicumstances. Table 1 gives a simple decision table for the solution of mathematical equations by analytical methods, ranging from the trivial to the impossible. Finally, there are considerable difficulties in the use of analytical models for the description of ecological processes which are discontinuous, as are many of the processes of plant and animal reproduction. Analytical models have not, therefore, been found to be of particular value in the study of ecosystem dynamics and in the modelling of whole ecosystems.

As a result, and because of the greater interest in modelling complex ecological processes, increased attention has been given to simulation models which use simpler forms of mathematics in conjunction with the speed and power of modern electronic computers. By placing any equation set on a computer and exploring interactively the consequences of changing the

Table 1 Mathematical problems and their ease of solution by analytical methods

Type and number of equations														
Linear	Y	Y	Y	Y	Y	Y	Y	Y	N	N	N	N	N	N
One equation	N	N	N	Y	Y	Y	Y	Y	N	N	N	Y	Y	Y
Many equations				N	N	N	Y	Y				N	N	Y
Algebraic	Y	N	N	Y	N	N	Y	N	Y	N	N	Y	N	
Ordinary differential		Y	N		Y	N				Y	N			
Partial differential			Y			Y					Y			
Analytical solution is:	↓	↓	↓	↓	↓	↓	↓	↓	↓	↓	↓	↓	↓	↓
Trivial	X													
Easy		X		X										
Difficult			X		X		X		X			X		
Very difficult										X				
Impossible						X		X			X		X	X

parameters of the equations, or the values taken by the variables, it is possible to obtain increased understanding and improved prediction of the dynamics of ecological systems and their components. Such simulation models do not usually give an exact solution to an equation over time as does an analytical model, and they are frequently subject to errors which are introduced by the inexact nature of the techniques of obtaining a solution. However, it is usually possible to solve many equations nearly simultaneously, and it is also possible to include a wide variety of non-linear relationships with the equations. The remainder of this chapter describes the use of simulation models of dynamic systems.

5.2 DEFINITIONS AND NOTATION

Some careful definition of terms and of mathematical notation is necessary for the understanding of dynamic models. Indeed, failure to conform to reasonable standards and definitions has done much to make dynamic modelling difficult to understand by the non-mathematician, and hence unpopular. In this section, we will therefore define some of the more important terms and their mathematical notation.

The variables which describe the state or condition of an ecological system are called *state variables*, and are usually shown as a capital letter qualified by a subscript e.g. Y_i, X_j, etc. A model which has more than one state variable is called multi-dimensional.

Rates of movement or energy between factors described by state variables of a system are called *flow rates*, and the symbol $M(i, j)$ denotes the flow rate of material from, say, state variable Y_i to Y_j. The symbol $M(i, i)$ represents that particular change in the state variable Y_i which is not a result of flows from other state variables. As an example, consider the simple predator–prey model in Figure 5. Y_1 in this figure is the population of some organism, and Y_2 is the population of another organism which acts as a predator on the first. Then $M(1,2)$ is the rate at which the prey Y_1 are being consumed by predators Y_2. Similarly, $M(1, 1)$ is the net growth (births minus deaths) of the prey population, and $M(2, 2)$ is the net growth of the predator population in the absence of predation. There is no $M(2, 1)$ as no material moves from the predator to the prey.

The flow rates $M(i, j)$ will usually be affected by factors other than those described by the state variables. Model variables whose values are determined by management decisions are called *decision variables*. Such variables are controlled solely by human decisions. Flow rates may also be influenced by factors outside the system, but which nevertheless affect the behaviour of the system. These variables are called *exogenous variables* or *driving variables*. In dynamic ecological models, driving variables describing environmental conditions, such as temperature, precipitation and solar radiation, often have an important influence on flow rates like primary productivity, decomposition or nutrient uptake. It is not unusual for a flow rate to depend on several state variables, and on several exogeneous or decision variables. Rates that depend on several factors are thus represented mathematically by adding together or multiplying together effects of factors that are assumed to act independently of one another.

The total *rate of change* of a state variable, Y_i, is the sum of effects due to flow into Y_i from other variables less the flows from Y_i to the other variables.

$$F_i = \sum_{j=1}^{N} \alpha_{ji} M(j, i) - \sum_{\substack{k=1 \\ k \neq i}}^{N} M(i, k)$$

where F_i is the total rate of change of a state variable Y_i; α_{ji} is a conversion ratio of material Y_i to material Y_j;

$\sum_{\substack{k=1 \\ k \neq i}}^{N}$ is the sum for $k = 1, 2, \ldots, i-1, i+1, \ldots, N$;

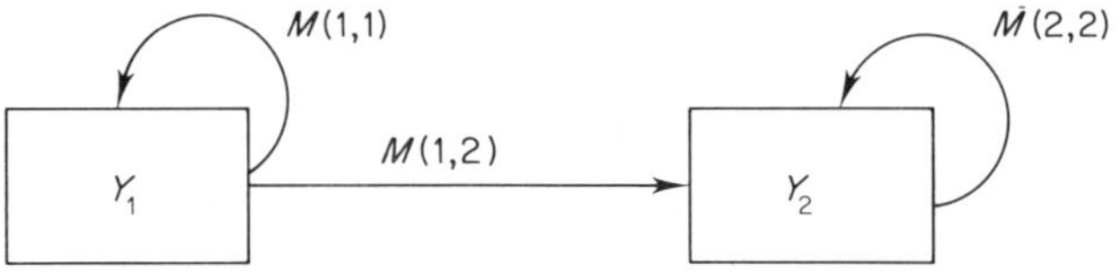

Figure 5. Simple example of state variables and flows in predator–prey model

and

$M(j, k)$ are the flow rates from Y_j to Y_k.

Rates of change are of fundamental importance in dynamic models because they are the basis of differential and difference equations used to predict the behaviour of ecological systems.

Some, or all, of the variables included in a model may be *random variables*, with each of which is associated a probability distribution that describes quantitatively the likelihood that the variable has a specified value, or falls within a specified range. Models which contain random variables are called *stochastic models*, and the output of such models represent probabilities rather than the fixed values which will be obtained from a *deterministic model*.

Finally, systems in which the variables change with time are called *dynamic*. Dynamic systems, and models of dynamic systems, may eventually reach a state at which the values of the variables do not change. A system, or model, in which all the state variables remain constant with time is said to be in a *steady state* or *equilibrium*. Long-term studies of systems may therefore concentrate on their steady-state behaviour rather than on transient fluctuations, and the calculation of the values of the state variables at which the system reaches a steady state is of particular importance.

5.3 DIFFERENTIAL AND DIFFERENCE EQUATIONS

Models of dynamic change incorporate rates of change, F_i with values of state variables, Y_i, so as to predict the altered states of those variables with time. If time is divided into discrete units such as days, generations or years, the value of the state variable, Y_i, at time $t + \Delta t$ is the value of the state variable Y_i, at time t, plus the rate of change, F_i, multiplied by Δt, i.e. the number of time units that have elapsed. In mathematical terms, therefore:

$$Y_i(t + \Delta t) = Y_i(t) + F_i(t)\ \Delta t$$

Such an equation is called a *difference equation*, and can be used to predict changes at discrete points in time.

In contrast, equations that describe the dynamics of a system continuously through time are called *differential equations*, and are based on the derivative of $Y_i(t)$. The *derivative* is a mathematical expression of the instantaneous rate of change of $Y_i(t)$, and is denoted by the function $\mathrm{d}Y_i(t)/\mathrm{d}t$, or alternatively by $\dot{Y}$.

$$\frac{\mathrm{d}Y_i}{\mathrm{d}t} = \lim_{\Delta t \to 0} \frac{Y_i(t + \Delta t) - Y_i(t)}{\Delta t}$$

Under defined mathematical conditions, this limit can be shown to exist and the function $Y_i(t)$ is said to be differentiable at the point t.

The purpose of predictive equations in dynamic modelling is to determine the values of Y_i at each time t. An expression that gives these values of $Y_i(t)$ is called the *solution* of the predictive equations. One of the principal advantages of differential equations is that a closed form or analytical solution may be possible from the mathematical expression of the relationships. For example, the exponential equation:

$$\frac{\delta Y}{\mathrm{d}t} = rY$$

where $Y(t_0) = c$, has the closed form solution:

$$Y(t) = \mathrm{c}e^{r(t-t_0)}$$

In general, however, closed form solutions are only attainable for models which are very simple, or which have some special form. Because of the complexity of most ecological systems, models of dynamic systems must be solved numerically, and the modern computer has greatly increased our ability to find numerical solutions to such problems.

The numerical solution of difference equations is found by recursive calculation of the values of the state variables, i.e. the values of the state variables are used to calculate the values at time $(t+1)$, which in turn are used to calculate the values at time $(t+2)$, etc. The process starts at $t = t_0$, for which initial values $Y_i(t_0)$ are given. Comparison of recursive and analytical solutions shows that the state variables are consistently underestimated by the use of difference equations. This underestimation is caused by the false assumption that the growth rate remains constant for the whole of the time period of a single step, when, in fact, the rate changes continuously. The discrepancy between recursive and analytical solutions decreases as the time interval is made smaller. One of the principal problems of obtaining solutions by the use of difference equations, therefore, is the choice of an appropriate time interval. If it is too short, the length of time needed for the solution increases dramatically; if it is too long, the underestimate of the state variable distorts the solution.

It may be argued that an acceptable time interval is one which gives an answer within a specified range from the answer obtained by an analytical solution. However, this is not a particularly helpful criterion. It is not necessary to find an approximate answer if an analytical solution exists, and, if no analytical solution can be found, we have no standard of comparison for the approximation. An alternative approach, therefore, is to ensure than an acceptable time interval has been reached when halving its value does not change the relevant results of the simulation by more than a pre-set amount. This pre-set amount should not, in general, be smaller than is warranted by the accuracy of the parameters, initial values and tabulated functions. In this way, a reasonable compromise may be achieved between the apparent accuracy of the answer and the computational effort, which increases at least linearly with the decreasing magnitude of the time interval.

More sophisticated methods of integration of differential equations are, however, frequently desirable, especially where it is necessary to vary the size of the time interval during the simulation, as frequently occurs in the more complex models. A wide range of methods exist for calculating the solutions to differential equations. Many of these methods are based on a Taylor series expansion of the function, $Y(t)$. From Taylor's theorem, any finite real-valued function, $Y(t)$, for which a finite derivative $\mathrm{d}Q/\mathrm{d}t$ exists over the interval $t = \mathrm{s} + \mathrm{t}$, can be represented by the expansion:

$$Y(t+\Delta t) = Y(t) + \frac{\mathrm{d}Y(t)}{\mathrm{d}t}\frac{\Delta t}{} + \cdots + \frac{\mathrm{d}^m Y(t)}{\mathrm{d}t^m}\frac{(\Delta t)^m}{m!} + \varepsilon(\Delta t^{m+1})$$

where m is any integer greater than zero and $\varepsilon(\Delta t^{m+1})$ is an error term less than a constant time $(\Delta t)^{m+1}$. Provided that $t < 1$, this error term becomes very small as m increases. If $m = 1$, then the expansion is reduced to:

$$Y(t+\Delta t) = Y(t) + \frac{\mathrm{d}Y(t)}{\mathrm{d}t}\Delta t + \varepsilon(\Delta t^2)$$

One of the simplest techniques for solving differential equations numerically is based on the reduced Taylor expansion, and is due to Euler. The value of the state variable Q at time $t + \Delta t$ is calculated from the equation:

$$Y(t+\Delta t) = Y(t) + \frac{\mathrm{d}Y(t)}{\mathrm{d}t}\Delta t = Y(t) + F(Y, t)\,\Delta t$$

where $F(Y, t)$ is the rate of change. The error of this approximation, $\varepsilon(\Delta t^2)$, will be small if t is less than 1. In essence, the solution becomes a difference equation that is used recursively to calculate the values of $Y(t)$ for $t_0 \leqslant t \leqslant T$ if $Y(t)$ is differentiable within the limits $t_0 < t < T$.

Euler's method requires at least $(T - t_0)/\Delta t$ recursive calculations in the interval $t_0 < t < T$, and, if Δt is small, the number of calculations may be large. Even though the error at each single step is small, the accumulation of errors over many steps may lead to a considerable loss of accuracy. A method which has a smaller error than $\varepsilon(\Delta t^2)$ at each step is known as the Runge–Kutta method. This uses a difference approximation of the form:

$$Y(t+\Delta t) = Y(t) + \sum_{i=1}^{N} C_i F(Q(t) + a_{i,t} + b_i) + \varepsilon(\Delta t)$$

where C_i, a_i and b_i are chosen so that the expansion is equivalent to that of the Taylor series. When $N = 4$, the commonly used Runge–Kutta approximation is:

$$Y(t+\Delta t) = Q(t) + \tfrac{1}{6}(m_1 + 2m_2 + 2m_3 + m_4)\Delta t$$

where

$$m_1 = F(y(t), t)$$
$$m_2 = F(Y(t) + \tfrac{1}{2} m_1 \Delta t, t + \tfrac{1}{2} \Delta t)$$
$$m_3 = F(Y(t) + \tfrac{1}{2} m_2 \Delta t, t + \tfrac{1}{2} \Delta t)$$
$$m_4 = F(Y(t) + m_3 \Delta t, t + \Delta t)$$

The approximation is a difference equation which can be used recursively, but which has an error at each step of the order $(\Delta t)^5$. A computer subroutine for solving differential equations by the Runge–Kutta method is therefore an essential part of the ecologist's toolkit for the modelling dynamic systems. An alternative and rather simpler method of mid-point integration also gives quite acceptable accuracy, and a computer subroutine for this method is also generally useful.

BASIC algorithms for the Euler, Runge–Kutta and mid-point methods are given in the Appendix to this Handbook.

The individual steps in constructing and solving a dynamic model may be summarized as follows.

(i) Define the state variables, exogenous variables and control variables of the system to be modelled.

(ii) Estimate, or determine experimentally, the flow rates $M(j, i)$ between the state variables.

(iii) Calculate the rate of change of each variable as the sum of flows into the variable minus the flows from the variable, i.e.

$$F_i = \sum_{j=1}^{N} \alpha_{ji} M(j, i) - \sum_{\substack{k=1 \\ k \neq i}}^{N} M(i, k)$$

(iv) Describe the dynamics of the system by substituting each rate of change, F_i, into a series of difference or differential equations.

(v) Find a numerical solution to the predictive equations of the dynamic model. If only the equilibrium solution is of interest, it can be obtained by solving the equation $F(Y) = 0$.

(vi) Test the sensitivity of the model to small changes in the initial values of the state variables, and of the exogeneous and control variables. Test also the sensitivity of the model to small changes in the estimated flow rates.

(vii) When the model has been shown to be reasonably robust to small changes in its basic parameters, the effects of various management policies may be tested by trial and error simulation, or by optimization methods.

5.4 MODEL TYPES AND APPLICATIONS

5.4.1 Population models

One obvious application of dynamic models is the prediction of populations of organisms in response to various environmental effects. Such applications have a long history, and many of these are referred to in the case studies in Section 5.6. Here, therefore, we will concentrate on some simple, but important, examples.

The simple model for the growth of a single species is the exponential growth equation:

$$\frac{dY}{dt} = F = rY$$

in which the rate of change F is equivalent to the number, mass or density of the population multiplied by a constant growth rate, r.

The exponential growth model has a closed form solution:

$$Y(t+1) = Y(t) + rY(t)$$

Table 2 gives the computed values of the growth of a hypothetical population with a mass of 1 g at t_0, and a growth rate of 0.1 g/hour. The first column of this table gives the time from zero to 10. The second column gives the exact solution to the differential equation, while the second, third and fourth, respectively, give the Euler, mid-point and Runge–Kutta solutions. The Euler method quickly leads to quite serious underestimates of the mass of the population, while both the mid-points and Runge–Kutta methods give acceptable estimates.

A more realistic model of population growth of a single species is given by

Table 2 Comparison of solutions of exponential growth equation $y = e^{0.1t}$

		Mass (g) predicted by methods of		
Time	Exact solution	Euler	Mid-point	Runge–Kutta
2	1.221	1.210	1.221	1.221
3	1.350	1.331	1.349	1.350
4	1.492	1.464	1.491	1.492
5	1.649	1.611	1.647	1.649
6	1.822	1.772	1.820	1.822
7	2.014	1.949	2.012	2.014
8	2.226	2.144	2.223	2.226
9	2.460	2.358	2.456	2.460
10	2.718	2.594	2.714	2.718

the logistic growth equation:

$$\frac{dY}{dt} = rY\frac{K - Y}{K} = F$$

where the rate of change is equal to the number, mass or density of the population multiplied by a constant growth rate, r, and by a factor $(K - Y)/K$ which approaches unity as Y approaches the value of K. Again, this differential equation has an exact or closed form solution:

$$Y(t) = \frac{K}{1 + ce^{-rt}}$$

where $c = \frac{K}{Y(0)} - 1$

The computed values of the growth of a hypothetical population with a mass of 0.45 g at t_0, a growth rate of 0.05 g/hour, and an upper value for K of 6 g are given in Table 3. Again, the Euler solution is less satisfactory as an approximation to the exact solution than either the mid-point or Runge–Kutta method.

For both the exponential and logistic growth models, exact or closed form solutions exist, but, for most population models, such solutions are either mathematically difficult or impossible. As a simple example of such a model,

Table 3 Comparison of solutions of logistic growth equation $y = K/(1 + ce^{-0.05t})$

		Mass (g) predicted by methods of		
Time	Exact solution	Euler	Mid-point	Runge–Kutta
0	0.45	0.45	0.45	0.45
10	0.71	0.66	0.70	0.71
20	1.08	0.95	1.07	1.08
30	1.60	1.35	1.57	1.60
40	2.25	1.87	2.22	2.25
50	2.98	2.52	2.95	2.98
60	3.72	3.25	3.69	3.72
70	4.37	3.99	4.35	4.37
80	4.89	4.66	4.87	4.89
90	5.28	5.18	5.26	5.28
100	5.54	5.54	5.52	5.54
110	5.71	5.75	5.69	5.71
120	5.82	5.87	5.81	5.82
130	5.89	5.93	5.88	5.89
140	5.93	5.97	5.92	5.93
150	5.96	5.98	5.95	5.96

consider the growth of a population of bacteria which is not limited by either space or food, but for which the relative growth rate is dependent on a fluctuating daily temperature (de Wit and Goudriaan, 1974). If the temperature is assumed to be prediced by the equation:

$$T = 20 + 10(\sin(2\pi X/24))\ ^\circ\mathrm{C}$$

where X is the number of hours, it will fluctuate between 10°C and 30°C every 24 hours. If, further, the relative growth rate is assumed to be related to temperature by the quadratic equation:

$$G = 0.00554 + 0.01019T - 0.0001009T^2$$

then the exponential growth model has a variable relative growth rate:

$$\frac{\mathrm{d}Y}{\mathrm{d}t} = GY$$

Table 4 Comparison of solutions for temperature-dependent exponential growth of a bacterial colony

Hours	Temperature (°C)	Relative growth rate	Predicted mass (g)		
			Euler	Mid-point	Runge–Kutta
2	25	0.19	1.4	1.4	1.4
4	29	0.20	1.9	2.1	2.0
6	30	0.21	2.8	3.1	3.0
8	29	0.20	4.1	4.7	4.5
10	25	0.19	5.9	7.0	6.8
12	20	0.16	8.2	9.9	9.9
14	15	0.12	11	13	14
16	11	0.10	14	16	17
18	10	0.01	16	20	21
20	11	0.10	19	23	25
22	15	0.12	23	29	31
24	20	0.16	30	39	39
26	25	0.19	41	54	54
28	29	0.20	58	80	78
30	30	0.21	84	121	127
32	29	0.20	122	183	177
34	25	0.19	176	270	266
36	20	0.16	245	382	386
38	15	0.12	324	507	530
40	11	0.10	404	362	681
42	10	0.09	483	757	828
44	11	0.10	571	905	986
46	15	0.12	695	1125	1198
48	20	0.16	892	1488	1537

Table 4 summarises the predicted growth of a population of bacteria with a starting mass of 1 g under these assumptions, based on Euler, mid-point, and Runge–Kutta solutions. Here, the simple Euler solution gives a substantial underestimate of the growth of the bacterial colony after about 12 hours. The mid-point and Runge–Kutta solutions are roughly comparable, within the limits of the basic assumptions.

All of the above models can be solved relatively easily by the use of the simple computer algorithms DIFFEU, DIFFEQ and DIFFER which are given in the Appendix. The only changes that need to be made in the programs are the substitution of the required function in the DEF statement at the beginning of the program. Similarly, one or more of the algorithms can be incorporated into a larger program to simulate the growth of a population.

More complex population models may require competition between species, or even between plants of the same species, to be taken into account. For example, the number of plants of each species is a mixed stand of an agricultural or forestry crop is determined at the time of planting. The competition between each crop species can then be expressed in terms of 'relative space', a dimensionless variable which characterises the effects of crowding on available root and foliage space, nutrients, sunlight and associated factors. The actual production of dry matter can then be obtained from the product of 'relative space' and the maximum possible yield for dense monocultures. All three quantities are therefore functions of time.

To simulate, for example, the growth and competition of barley and oats planted as a mixed stand, the differential equations for the relative space r_b and r_o for barley and oats respectively are as follows:

$$dr_b = G_b(r_b - r_b^2 - r_b r_o)$$
$$dr_o = G_o(r_o - r_o^2 - r_b r_o)$$

where G_b and G_o are relative growth rates of barley and oats as empirical functions of time in the absence of competition summarized in Table 5.

Table 5 Empirical growth rates for barley and oats

	Relative growth rate	
Time in days	Oats	Barley
0	0.43	0.71
7	0.11	0.12
14	0.04	0.06
21	0.02	0.04
28	0.006	0.02
35	−0.004	0.05
42	−0.007	0.05

Table 6 Computed proportions of growing space for oats and barley growing in competition

Time in days	Proportion of growing space	
	Oats	Barley
1	0.273	0.468
14	0.312	0.523
21	0.329	0.542
28	0.339	0.549
35	0.349	0.549
42	0.361	0.547

The calculated proportions of the growing space occupied by the two species when sown at an initial spacing of 0.02 cm are given in Table 6. Because the barley maintains its relatively fast growth rate, a disproportionate share of the available space is occupied by barley at an early stage when the two species are grown together. By the time the oats begin to grow, there is insufficient space to accommodate them.

5.4.2 Predator–prey models

While population models, with or without competition or the constraints on resources to support growth, represent one of the main areas of application for dynamic differential or difference equation models, one of the most popular areas for such models has always been that of the predator–prey relationship. In this simple review of predator–prey relationships, we will first consider predator–prey systems without an age structure and then extend the models to include systems which have an age structure.

5.4.2.1 Predator–prey species without an age structure

Dynamic models of predator–prey interaction are usually based on several simplifying assumptions, including the following.

(i) The density of species, usually expressed as the number of individuals per unit area, can be adequately represented by a single variable. Differences of age, sex, genotype, etc., are all assumed to be capable of being ignored for the purpose of this model.

(ii) Changes in the density of the organisms are assumed to be adequately described by deterministic equations, i.e. there are no obvious random or probabilistic events.

(iii) The effects of the interactions between the organisms are assumed to be instantaneous, i.e. there is no lag between the consumption of one organism by another and the response of the consuming organism to the increase of available energy.

If these relationships are true, or approximately true, then the predator–prey relationships may be capable of being represented by differential or difference equations.

5.4.2.2 *Volterra equations*

Volterra (1926) described the interactions between a prey species, x, and its predator, y, by the following equations:

$$\dot{x} = ax - bx^2 - cxy$$

$$\dot{y} = -ey + c'xy$$

We have already considered the analytical solution to these equations in Section 5.1, and suggested that the system converges to a steady state with prey and predator both present only when the carrying capacity a/b is high enough to support the predator (Maynard-Smith, 1974). However, it will be interesting to explore this simple representation of a predator–prey relationship by the same methods described for population models. If we can show that these methods give the same results as the analytical solutions, where these exist, we may have more confidence in the use of the numerical methods in situations where the analytical solutions do not exist.

Analytical solution of these equations showed that both predator and prey numbers oscillate with decreasing amplitude, the predator oscillations lagging in phase behind the prey. If $y = 0$, the prey species is limited only by the predator, and the prey increases exponentially in the absence of the predator. The oscillations are then of constant amplitude, depending on the initial conditions — a system started close to its steady state will have small-amplitude oscillations, and one started far from its steady state will have large-amplitude oscillations. Such a system is called 'conservative' because there is a quality which is conserved during the motion, as energy is conserved in simple harmonic motion.

The term bx^2, expressing the inhibiting effect of a species on its own growth, is referred to as a 'damping' term. In ecology, the main factor reducing oscillations is the presence of self-inhibiting effects. The Volterra equations, without damping, are written as:

$$\dot{x} = ax - cxy$$

$$\dot{y} = ey + c'xy$$

From these equations can then be derived the equation:

$$e\ln(x) - c'x + a\ln(y) = \text{constant}$$

which represents a family of closed curves in which each member of the family corresponds to a different value of the constant. Choice of a starting point, i.e. of initial values of x and y, determines the value of the constant. Any population will continue indefinitely to follow the cycle on which it starts. When the cycles are of very small amplitude, they may be approximated by ellipses. For simple models of the kind illustrated by the Volterra equations, it is possible to derive the equation of such an ellipse, and some interesting properties can be inferred from it.

Four properties are true of small cycles (in the neighbourhood of the equilibrium point).

(i) The sizes of both the prey and the predator populations vary sinusoidally with period $T = 2\pi/ae$. The period is the same for both species and depends only on the parameters a and e.
(ii) The two populations are always a quarter-cycle out of phase. Thus, the prey population begins to decrease from its maximum size at the instant when the predator population is attaining its fastest growth. One quarter-cycle later, when the prey population is declining most rapidly, the predator population attains its maximum size and starts to decline.
(iii) The amplitudes of the oscillations, unlike the period of the oscillations, depend on the initial sizes of the populations.
(iv) The mean size of the prey population over a period T is e/c', and its identical to the equilibrium size of the prey population. Similarly, the mean size of the predator population is a/c.

Consider a difference equation model of a predator and prey equation whose rate of change is defined by the equations:

$$F_1 = r_1 Y_1 - a_1 Y_1 Y_2$$
$$F_2 = r_2 Y_2 + a_2 Y_1 Y_2$$

Then:

$$Y_1(t+1) = Y_1(t) + r_1 Y_1(t) - a_1 Y_1(t) Y_2(t)$$
$$Y_2(t+1) = Y_2(t) + r_2 Q Y_2(t) - a_2 Y_1(t) Y_2(t)$$

If $r_1 = 0.5$, $a_1 = a_2 = 0.001$, and $r_2 = -1$, and $Y_2 = -1$, and $Y_1(0) = 1000$ and $Y_2(0) = 400$, recursive calculation of these difference equations gives the results shown in Table 7.

The properties of dynamic models of this kind and readily explored by computers, especially when provided with graphics terminals. Figure 6 gives a simple difference equation solution, using the Euler approximation, derived from the BASIC algorithm in the Appendix. In this program, the solution is given for the equations:

$$x = 4x - 2xy$$
$$y = 3y + xy$$

Table 7

Time	Y_1 Prey	Y_2 Predator
0	1000	400
1	1100	400
2	1210	440
3	1283	532
4	1241	683
5	1014	847
6	662	859
7	424	569
8	395	241
9	497	95
10	698	47

with the starting values of $x = 3, y = 0.5$; $x = 3, y = 1$; $x = 3, y = 1.5$; and $x = 3, y = 2$.

The results in Figure 6 show three roughly ellipsoid trajectories and a single point, this point corresponding to the starting value of $x = 3, y = 2$ and representing equilibrium. If the two populations start at this point, no change ever occurs. For any other starting values, the populations of both prey and predator circle about the equilibrium point in a counter-clockwise direction.

A modification of this algorithm shows the effect of the damping term $-bx^2$, by changing statement 120 of the program to

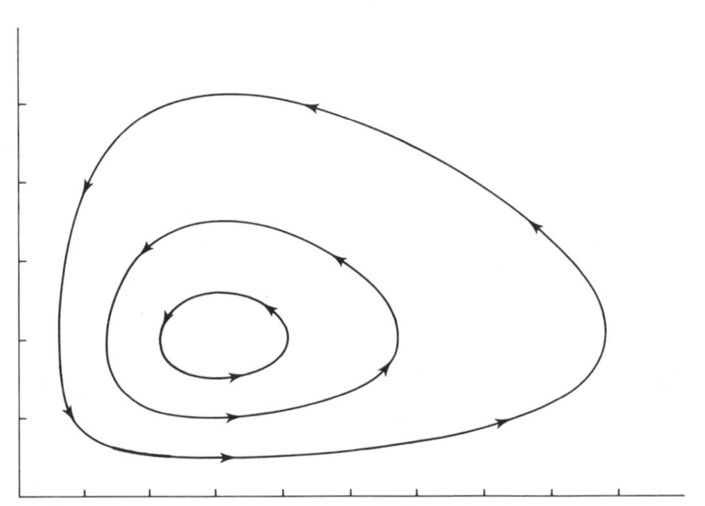

Figure 6. Trajectories of predator–prey model for four different starting points

$X = X + (4 * X - 0.25 * X * X - 2 * X * Y) * 0.01$. If the starting values are set to $x = 3, y = 0.5$, the sizes of the two populations quickly lead to the equilibrium point of the solution (Figure 7).

Use of computer models and difference equations in this way, even with relatively inefficient methods such as the Euler approximation, can therefore be helpful in showing the behaviour of systems of equations. They are particularly useful when, in contrast to the simple model of the Volterra equations, analytical solutions cannot be readily derived. A general-purpose algorithm for plotting the trajectory of predator–prey populations is given as the program PREDAT in the Appendix.

Leslie (1948) gave an alternative formulation of the predator–prey equations:

$$\dot{x} = ax - bx^2 - cxy$$
$$\dot{y} = ey - fy^2/x$$

The equation for the prey is the same as Volterra's equation with damping. The equation for the predator resembles the logistic equation, but the second term has been modified to allow for the density of the prey. If x/y is large (many prey per predator), the predator increases exponentially. If $x/y = f/e$, the predator is at equilibrium. If x/y is less than f/e, the predator decreases in numbers. The equations lead to rapidly damped oscillations.

Volterra's equations are usually to be preferred for two reasons.

(i) In the Volterra equations, whether the predator increases or decreases depends only on the density of the prey, while, for the Leslie equations, it depends on the number of prey per predator.

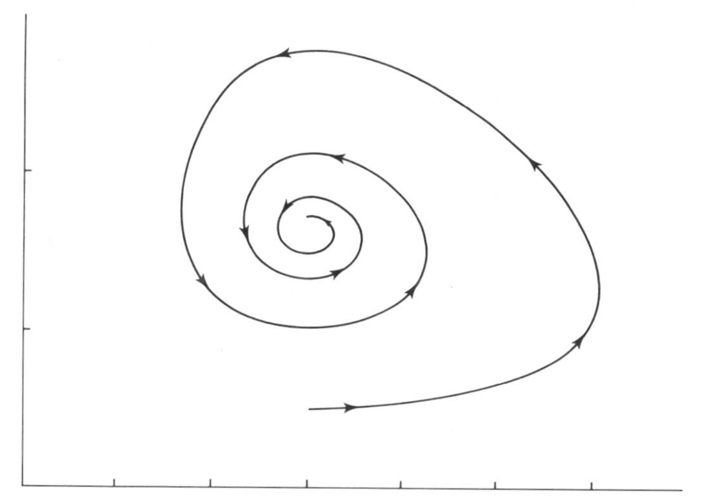

Figure 7. Trajectory of predator–prey model when damping term is included

(ii) The Volterra equations relate the rate of increase of the predators to the rate ($c'xy$) at which the prey is being eaten. In the Leslie equations, there is no relationship between the rate at which a predator eats and the rate at which it reproduces.

5.4.2.3 *The effect of protection for the prey*

Suppose that some number, x_r, of the prey can be protected so as to make them inaccessible to the predator. Volterra's equations (without damping) then becomes:

$$\dot{x} = ax - cy(cx - x_r)$$
$$\dot{y} = ey + c'y(x - x_r)$$

If the number of prey which are protected is a constant fraction of the total, so that $x_r = kx$, the equations can be written as:

$$\dot{x} = ax - c(1-k)xy$$
$$\dot{y} = -ey + c'(1-k)xy$$

Such a change clearly does alter the nature of the equilibrium. If, however, the number of the prey which are protected is constant, so that $x_r = k$, then it can be readily shown that the effects of the protection are stabilizing, changing a 'conservative' system into one with convergent oscillation.

5.4.2.4 *Predator with constant food intake*

For some predators, a more reasonable assumption that an individual predator takes prey at a constant rate is determined by the food requirements of the predator and not by the density of the prey. The equations then become:

$$\dot{x} = (a-b)x - cy$$
$$\dot{y} = -cy + c'y = (c' - e)y$$

However, the equation for the predator is unrealistic, as it provides for an exponential increase if $c' > e$, or an exponential decrease if $c' < e$. If we assume that a predator has a constant food intake, independent of prey density, the predator cannot be limited by its prey. The predator must, therefore, be self-limited in some way, with a damping term in the equation for y, i.e.:

$$\dot{y} = (c' - e)y - dy^2$$

Such a system has two stationary points when x and y are non-zero. One of these points is a stable non-oscillatory equilibrium and the other is an unstable equilibrium. Which of the two stationary points the system will reach depends on the initial conditions.

The model is unrealistic for any initial conditions which are not close to the stable equilibrium point, because the assumption that every predator takes prey at a constant rate, regardless of prey density, must ultimately break down when the prey become rare.

5.4.2.5 General case — Rozenzweig–MacArthur model

A more general case is given by the equations:

$$\dot{x} = f(x) - \phi(x, y)$$
$$\dot{y} = -ey + k\phi(x, t)$$

where $f(x)$ is the rate of change of x in the absence of predators;
$\phi(x, y)$ is the rate of predation;
k is the conversion efficiency of prey into predator; and
e is the mortality rate of the predator.

Analysis of these equations for continuously reproducing systems gives the following conclusions.

(i) The interaction of predators and prey, when the predator is limited only by the supply of prey, leads to regular fluctuations in the numbers of both predators and prey.
(ii) If the numbers of prey are limited by resources, and not by the predator, the oscillations will tend to be damped out.
(iii) If the predator is limited by some other factor than the prey, the oscillations will also tend to be damped out.
(iv) If some form of protection makes a constant number of prey unavailable to the predators, the oscillations will again tend to be damped out.
(v) Fluctuations are likely to increase in amplitude, perhaps leading to the extinction of one or both species, if the predator is able to maintain itself when the prey density falls below the level necessary for the survival of the predator.

5.4.2.6 Stochastic simulation of predator–prey populations

A stochastic version of the purely deterministic Volterra equations:

$$\dot{x} = ax - bxy$$
$$\dot{y} = cy + dxy$$

can also be simulated, assuming that these equations represent the next outcome of a large number of random events.

Suppose that a is the average birth rate for the prey, and that all prey deaths are due to attack by the predators. Similarly, we suppose that c is the average death rate of the predator, and that the birth of a predator always coincides with the death of a prey animal, e.g. if the predators are parasitic wasps that lay their eggs in the larvae of host insects so that development of each parasite necessarily involves destrucion of host larvae or pupae. With this limitation, the number of possible events is either three of four, the case with three possible events arising when every death of a prey animal is accompanied by the birth of a predator, so that $b = d = p$ (say). The events, and their associated probabilities, are summarized below, where c is a constant.

Event	*Probability*
$x \to x + 1;\; y \to y$	cax
$x \to x;\; y \to y - 1$	ccy
$x \to x;\; y \to +1$	$cdxy$

If some prey deaths do not coincide with predator births, and only a proportion of predator attacks are successful, then $d = \theta b$, where θ is the fixed proportion. The event $x \to x - 1$ has two possible outcomes: either $y \to y + 1$ with probability proportional to $\theta bxy = dxy$, or $y \to y$ with probability proportional to $(1 - \theta)bxy = (b - d)xy$. The case $\theta = 1$ is the special case above.

The risk of extinction of one or the other of the two species is clearly greater when the trajectory of the combined populations passes close to either of the axes. When a population reaches a point in its trajectory such that one of the species has only a few members, the probability that species will fail to recover is high. If the predator dies out, the prey population will grow without restriction until some density-dependent or regulating function comes into operation. If the prey dies, the predator must also die, unless it can transfer its dependence to some other species.

5.4.2.7 *Predator–prey systems with age structure*

So far, we have considered predator–prey systems with no defined age structure, thus keeping the mathematical representation of the systems as simple as possible. For most ecological systems, however, the effects of breeding seasons and age structure cannot be ignored. These effects often cause oscillations of large amplitude to occur in the behaviour of the systems through the operation of the phenomenon which is known to engineers as 'feedback'. As a general rule, if the duration of the delay in a feedback loop is longer than the 'natural period' of a system, oscillations of large amplitude will result. The 'natural period' for the system is readily defined as follows.

If the growth of a population in the absence of a feedback loop governing its regulation is represented by the differential equation:

$$\frac{\mathrm{d}x}{\mathrm{d}t} = rx$$

then the natural period is $1/r$.

5.4.2.8 *Delayed regulation of systems*

Delay in the regulation of ecosystems may arise from one or more of three causes.

(i) *Development time.* Changes in the environment of a species may produce an immediate change in (say) the rate at which eggs are produced by mature females. However, the corresponding change in the number of adults will be delayed by the length of time that it takes for an egg to develop into an adult, i.e.:

$$\mathrm{d}x/\mathrm{d}t = f(Xt - t_0)$$

where $Xt - T$ is the adult breeding population at some time t_0 in the past.

(ii) *Discrete breeding seasons.* If a species breeds only at a particular time, e.g. only in the spring of each year, some delay is necessarily introduced into the processes which regulate the ecosystem, even where the individual organisms usually survive in breeding again in successive years. Where the individuals of a species live for many years, and produce relatively few young each year, the delay of one year is likely to be short compared to the natural period for the species, so that any oscillations in numbers are convergent. Where, however, the adults breeding in one season survive only rarely to breed in the next year, the equation becomes:

$$x_{n+1} = \phi x_n$$

where x_n is the population size in year n. Even where the differential equation has a stable equilibrium, the corresponding difference equation will show divergent oscillations.

(iii) *Delayed response by limiting factors.* The numbers of a species may oscillate if there is a delay in the response to the factors limiting the population size of the species, even if the species itself responds quickly to the environmental conditions.

Maynard-Smith (1974) gives some simple examples of predator–prey systems incorporating such delays.

5.4.2.9 *Matrix representation of population growth with age-dependent birth and death rates*

Where a population can be divided naturally into a series of discrete age classes, a matrix representation of the population has been suggested by Leslie (1945, 1948). At any given time t, the population can be represented by a column vector:

$$\begin{bmatrix} n_{0t} \\ n_{1t} \\ n_{2t} \\ \vdots \\ n_{mt} \end{bmatrix}$$

In this vector, n_{it} is the number of females of age i at time t.

The population structure at the next time period $t+1$ is then given by equation:

$$\begin{bmatrix} n_{0,t+1} \\ n_{1,t+1} \\ n_{2,t+1} \\ \vdots \\ n_{m,t+1} \end{bmatrix} = \begin{bmatrix} F_0 & F_1 & \dots & F_{m-1} \\ P_0 & 0 & \dots & 0 \\ 0 & P_1 & \dots & 0 \\ & & \vdots & \\ 0 & 0 & \dots & P_{m-1} \end{bmatrix} \begin{bmatrix} F_m & n_{0t} \\ 0 & n_{1t} \\ 0 & n_{2t} \\ & \vdots \\ 0 & n_{mt} \end{bmatrix}$$

In this equation, F_0, F_1, F_{m-1}, F_m are the fecundities of females of different ages, i.e. F_i is the number of daughters born to a female of age i which survive to the next time interval, and so contribute to the number $n_{0,t+1}$. Similarly, P_i is the probability that a female of age i will survive to age $i+1$, i.e. $n_{i+1,t+1} = P_i n_{it}$.

The matrix notation is particularly convenient for the purposes of calculation and for the analysis of the basic properties of ecological systems (Pielou, 1969; Williamson, 1972). Usher (1972) gives an extensive account of the developments of the Leslie matrix model for a wide range of applications, including predator–prey systems.

5.4.3 Energy flow and nutrient cycles

In ecosystem research, we are often concerned with such processes as the cycling of nutrients and the flow of energy. Dynamic models can be well adapted so as to deal with such processes, representing the input of energy or nutrients into the ecosystem, and the transfer of energy or nutrients within the ecosystem. The natural compartmentation of the ecosystem into its species composition or into its trophic levels facilitates the formation of equations describing the distribution of energy or nutrients in the ecosystem at any given time (Phillipson, 1966). Losses from the ecosystem may be specified explicitly, or they may be assumed to be the difference between input and the sum of the output from, the storage in, any one compartment. Initially, it is simpler to assume that the method of creating compartments for an ecosystem is through trophic levels.

Figure 8 gives a simple example of the flow of phosphorus in a three-compartment system taken from Smith (1970). The parameters of this model

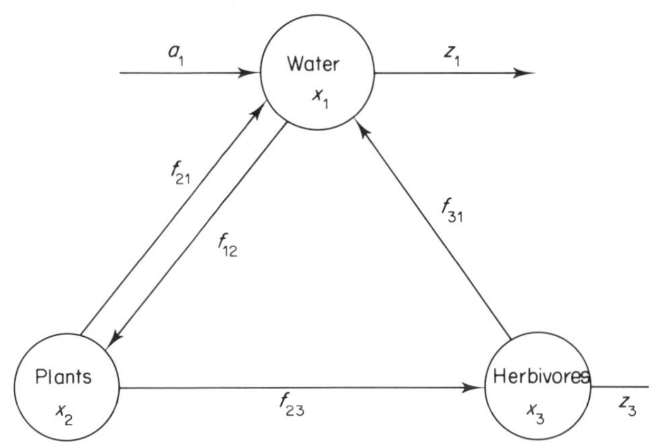

Figure 8. Representation of the phosphorus cycle in a three-compartment ecosystem

are defined as:

x_i = the amount of phosphorus in the ith compartment at any specified time;
a_i = the rate of inflow of phosphorus into the ith compartment;
f_{ij} = the rate of flow of phosphorus from the ith to the jth compartment;
f_{ii} = the proportion of phosphorus that is stored in the ith compartment during a given period of time.
z = the rate of outflow of phosphorus from the ith compartment.

In matrix notation, the equations for this model are given in Figure 9. The two vectors in this equation denote $x_{t,i}$ of energy or nutrients in the ith compartment at time t. The matrix elements of the form $f_{ij}(i \neq j)$ denote the transitions between the ith and jth compartments. The elements in the leading diagonal are composed of two factors, namely the energy or nutrient not transferring between compartments (f_i) and the input to the ith compartment, which is, in this model, independent of the amount of energy or nutrients already in that compartment, and is therefore represented as $a_i/x_{t,i}$. For the parameters summarized in Table 8, it can be shown that this ecosystem is in a steady state, with no increase in the total amount of phosphate.

Usher (1972) summarizes methods of analysis appropriate to matrix models of energy and nutrient cycles, and gives an example of the energy flow in a four-compartment ecosystem. Alternatively, the algorithm in the Appendix can be used to explore the changes in the state variables.

$$\begin{bmatrix} f_{11} + a_1/x_{t,1} & f_{21} & f_{31} \\ f_{12} & f_{22} + a_2/x_{t,2} & f_{32} \\ f_{13} & f_{23} & f_{33} + a_3/x_{t,3} \end{bmatrix} \begin{bmatrix} x_{t,1} \\ x_{t,2} \\ x_{t+1,3} \end{bmatrix} = \begin{bmatrix} x_{t+1,1} \\ x_{t+1,2} \\ x_{t+1,3} \end{bmatrix}$$

Figure 9. Matrix representation of phosphorus cycle

Table 8 Parameter values for three-compartment ecosystem

Parameters	Initial values
x_1	9.5
x_2	1.4
x_3	9.0
f_{12}	1.39
f_{21}	0.073
f_{23}	1.22
f_{31}	0.47
a_1	1.04
z_1	0.197
z_3	0.80

5.5 COMPUTATIONAL SOLUTIONS OF DYNAMIC MODELS

Scientists with a sound knowledge of computer programming can quite easily write efficient algorithms for the computational solution of dynamic models. With the development of high-level computing languages like FORTRAN or BASIC, it is a relatively simple matter to write down the difference equations describing the effects of flows on the system variables, and subroutines can be used recursively in order to obtain the solution of dynamic models, and also to explore their consequences for the management of natural systems. Some of these simple algorithms are given in this Handbook.

However, as interest has grown in the use of dynamic models in a wide range of applications, special-purpose languages have been developed to simplify the task of programming the computers used in deriving the solutions. One of the earliest of these languages was CSMP, standing for Continuous System Modeling Program. CSMP is a superset of the general-purpose language FORTRAN, and, as a result, is generally well maintained and supported by computer centres. The practical advantages of the language, and of other similar languages, is that the rate equations can be written down in a relatively simple way, and without any regard to the order in which they must be calculated. Given the necessary parameters for the model and the time limits within which the solutions are required, the translation of the model into the underlying FORTRAN is performed automatically. An added advantage of CSMP, in particular, is that it contains a wide choice of approximate integrating procedures which can be selected by the modeller, depending on the nature and purpose of the model. This opportunity to choose can be a most attractive characteristic in modelling biological systems, with their frequently complex and interacting continuous relationships. Like most languages of this type, it incorporates automatic time-keeping routines and sophisticated output facilities, including the production of graphs and diagrams. Anyone wanting to use CSMP for the modelling of dynamic systems should consult their local computing centre. The books by de Wit and Goudriaan (1974), Dent and Blackie (1979) and Brockington (1979) all contain examples of CSMP programs.

A perhaps even more widely used integrating simulation language is DYNAMO (an acronym for DYNAmic MOdels), which is available for most mainframe and microcomputers. Although DYNAMO is similar in appearance to CSMP, it has a more limited choice of integrating procedures. In contrast, DYNAMO provides excellent facilities for the exploration of models. This exploration is achieved by examining a whole series of specific simulations of the model with variations in parameter values, exogeneous data, model structure, etc., and provides an insight into overall system performance. Again, anyone wishing to use DYNAMO should consult their local computer centre. An excellent introduction to computer simulation through the use of DYNAMO is given by Roberts *et al.* (1983).

With the rapid development of microprocessors, improved special languages for the modelling of dynamic processes can be expected to be developed, exploiting, especially, the high-resolution graphics that have become almost commonplace on such machines. On the other hand, the interactive facilities provided by microprocessors for both model development and model exploration make it even more desirable than before that scientists learn to program computers for themselves. New general- and special-purpose languages are constantly being devised which will make it possible for dynamic processes to be modelled and explored.

5.6 CASE STUDIES IN DYNAMIC MODELLING

As dynamic modelling is by far the most popular technique for modelling change in ecosystems, there are numerous case studies — certainly too many to enumerate in full. For the interested reader, however, there are some useful collections, and notably the following.

Hall and Day (1977) have written introductory chapters on modelling theory and procedures, followed by case studies using models to analyse natural systems, assess environmental impact, and design optimal (or, at least, better) interactions between man and nature. Four of the introductory chapters are of particular value. They begin with definitions of a system and systems analysis and then go through some basic steps generally used to develop conceptual, diagrammatic, mathematical and computer models. The presentation reflects the authors' belief that the mechanics of building a model are relatively simple, particularly when compared with the complexity of determining the important attributes and realations within the ecosystem. Hall *et al.* (1977) describe a symbolic language intended to assist in the conveyance of complicated and voluminous ecological information and in the analysis of complex systems. The energy flow language is based on a series of modules, but represents both system processes and mathematical functions, connected by lines representing transfer particles of energy, materials, or information. Overton (1977) emphasizes the structure and strategic aspects of the model building process, with little attention to tactics. The tactical matters of solving equations and choosing the best form for an explicit relationship are technical problems, but the availability of suitable tactics limits the possible strategies, and the selection of a strategy constrains the tactics so that the two aspects are closely interrelated. Finally, Shoemaker (1977a, b) describes the fundamental mathematical steps in the construction and solutions of models of an ecological system intended to understand and predict the behaviour of the system as a whole by quantitatively describing interactions between parts of the system. The discussion of mathematical models is limited to those models designed to use data as well as ecological principles to predict the behaviour of specific ecological systems.

Of the remaining papers describing particular case studies, one of the most important is that by Botkin (1977) which describes a computer model of forest growth, designed to simulate the growth of individual trees on small forest plots. Experience with the model suggests that is reproduces the major dynamic characteristics of a forest community, and that it can be used to construct simulated experiments about the response of a forest to perturbations and manipulation. These experiments given insight into the importance of species interactions to the success and survival of any one species, and to the persistence of the entire forest community. Further details of this model are given by Botkin *et al.* (1972a, b).

Jeffers (1972) contains the proceedings of a symposium on the use of mathematical models in ecology and related sciences. The volume contains some important papers, ranging from the philosophical discussion of various aspects of mathematical modelling in empirical science by Skellam (1972) to the detailed description of developments of the Leslie matrix model by Usher (1972) with its extensive bibliography, already quoted extensively in this Handbook. Goodall (1972) discusses the building of dynamic models of ecosystems for predictive purposes in terms of the identification of state variables, the processes involved in their changes and factors influencing their rate of change. Conway and Murdie (1972) illustrate the use of population models as an aid to pest control at several different levels, with examples drawn from models of component processes, including reproduction and predator–prey interactions, and population models of mosquitoes and red bollworm. Other papers emphasize the importance of parameter sensitivity (Plinston, 1972), stochastic model fitting by evolutionary operation (Ross, 1972) and the use of problem-orientated packages in the formulation of ecological models (Radford, 1972), in addition to a wide variety of applications to both large-scale and small-scale ecological problems.

Perhaps the most extensive set of case studies, however, occurs in the four volumes entitled *Systems Analysis and Simulation in Ecology*, edited by B. C. Patten. Patten's primer for ecological modelling and simulation with analogue and digital computers remains one of the best introductions to dynamic modelling, and gives a sufficient treatment of programming elements to permit ecological models of no small significance (systems of coupled differential equations) to be implemented effectively for simulation or systems analysis studies on both digital and analogue computers (Patten, 1971). Other chapters in the same volume, for example O'Neill (1971), apply these techniques to actual prediction of ecological phenomena, with varying success.

The second volume of the series (Patten, 1972) contains chapters describing applications of selected dynamic analysis procedures, in addition to a section on ecological systems theory, and an extensive final section on applications of social relevance. The theoretical discussions include, for the first time, such topics as the steady-state equilibrium in simple non-linear food webs, the structural properties of food webs, and the concept of niche pattern. The third volume (Patten, 1975) is broadly divided between a discussion of ecosystem modelling in the US International Biological Programme, with chapters on grassland, eastern deciduous forest and desert, tundra and coniferous forest biomes, and models of freshwater-estuarine ecosystems. The case studies contain considerably more detail than those in earlier volumes, consistent with the general development of the dynamic modelling techniques.

The fourth volume of the series (Patten, 1976) is divided into two main sections, the first dealing with models of estuarine marine ecosystems, terresterial ecosystems, and human ecosystems, and the second dealing with the special problems and theory of ecosystem modelling. The four models together provide a succinct review of the development of dynamic modelling of ecosystems over the most active period of this development, and the papers contained in these volumes provide an extensive bibliography of ecosystem modelling. The final volume, in particular, amply demonstrates the current scope of systems ecology from its past and present emphasis on the parts and mechanisms in simulation modelling, and its movement towards systems analysis and novel, more formal consideration of ecosystem theory. The seventeen chapters make clear that, although the systems approach is young in ecology, it has substantially enriched the science, both methodologically and conceptually.

One of the largest collections of mathematical models of air pollution, water pollution, ecology and other environmental topics is the proceedings volume of a conference on environmental modelling simulation held in Cincinnati, Ohio, in April 1976. Edited by W. R. Ott (1976), this volume contains 164 papers, covering a very wide range of environmental concerns. Many of these papers are too short to give adequate detail of the formulation and use of models in the environmental sciences, but, collectively, they provide a very extensive bibliography to guide further reading and investigation within highly specialized fields of application.

Jorgensen (1979a) also contains the proceedings of a conference, this time on the state-of-the-art in ecology modelling, held in Copenhagen in 1978. The first part of this book contains invited state-of-the-art reviews of river models (Hahn and Eppler, 1979), graphical methods (Maguire, 1979), microcosms (Lassiter, 1979), predator–prey (Dubois, 1979), water quality and irrigation (Lassiter *et al.*, 1979; Skogerboe *et al.*, 1979; Jorgensen, 1979b). The second part contains 29 original papers presented at the conference. Again, such a volume provides a review of present knowledge, and an introduction to the relevant specialist literature.

Shugart and O'Neill (1979) also provide a series of benchmark papers, i.e. studies which are thought to have had the greatest influence on later research. The authors have selected some early papers that had a significant influence, as well as more recent papers that represent innovative approaches. They have also included several papers that contain information or points of view that are important for the understanding of systems ecology.

Innis and O'Neill (1979) address some of the problems of constructing and analysing ecosystem models, and present the papers delivered on this topic at the Satellite Programme in Statistical Ecology held at various locations in 1976–1978.

Finally, for the reader interested in further exploration of the systems ecology literature, and particularly of models available only as internal reports, the bibliography of O'Neill *et al.* (1977) contains over 900 references to the modelling literature.

CHAPTER 6

Markov Models and Related Procedures

6.1 MARKOV PROCESSES

Many ecological processes exhibit a great deal of variability, but are nevertheless influenced, if not controlled, by events which have gone before. A Russian mathematician, Markov, who lived early in the 20th century defined one such class of processes, in which the probability of the process being in a given state at a particular time is related to the immediately preceding state of that process. A Markov chain is, therefore, a sequence or chain of discrete states in time or space with fixed probabilities for the transition from one state to a given state in the next step in the chain. In its simplest form, a Markov chain may be regarded as a series of transitions between different states, such that the probabilities associated with each transition depend only on the immediately preceding state, and not on how the process arrived at that state. Such a chain contains a finite number of states, and the probabilities associated with the transitions between the states do not change with time, i.e. they are stationary. A first-order Markov chain takes account of only a single step in the process, but the definition may be extended so that the probabilities associated with each transition depend on events earlier than the immediately preceding one. Furthermore, the Markov chain may exhibit multiple dependence relationships, so that the probabilities associated with each transition depend jointly on more than one previous event.

The mathematical model of the Markov chain occupies an intermediate position in the spectrum of dynamic models, ranging from the classical, deterministic models at one extreme to stochastic models with independent events at the other. In the Markov model a random component is present, so that the state of the system at any point of time or space is not wholly dependent on the previous event or events, but there is, nevertheless, a structure of successive events which defines the process. Many ecological processes can be shown to exhibit the Markov property, i.e. a dependence of the probabilities associated with each transition on the immediately preceding state (or states). Some of these processes can be described by the simplest form of Markov chain (first-order chains with stationary transition probabilities), but many require more complex relationships involving multi-dependence chains with non-stationary probabilities.

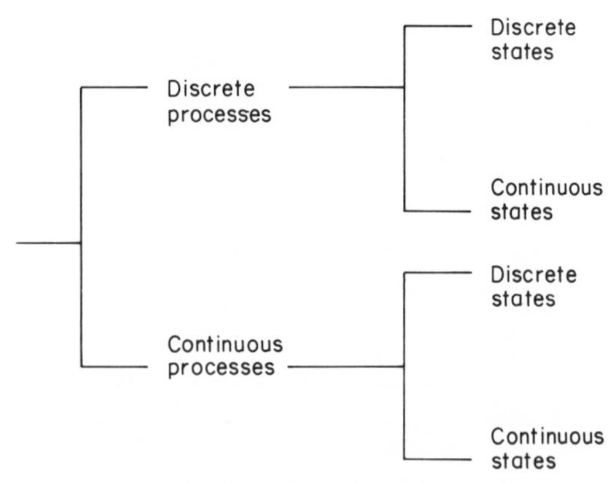

Figure 10. Classification of Markov processes

A simple classification of Markov processes distinguishes between processes in discrete and continuous time, and between states which are either discrete or continuous, as illustrated in Figure 10.

In this account of Markov models, most attention will be given to discrete-time, discrete-state processes, but continuous-time Markov processes are dealt with briefly later.

6.2 TRANSITION MATRICES

The matrix of transition probabilities provides a compact and unique description of the behaviour of a Markov chain. Each element in the matrix represents the probability of the transition from a particular state (represented by the row of the matrix) to the next state (representing the column of the matrix). Assuming a fixed number of possible states, the transition to and from every state can be described by a single matrix. A Markov transition matrix with three states, S_1, S_2, and S_3, is given (Figure 11), where each P_{ij} represents the probability of the transition from state S_i to state S_j. Because all events from any one state must either remain in the same state or move to one of·the others, the sum of the probabilities in each row is exactly 1.0. There is, however, no necessity for the sum of the probabilities in the columns to equal any fixed value.

$$\begin{array}{c} \\ \\ \textit{Present state} \end{array} \begin{array}{c} \textit{Future state} \\ \begin{array}{cccc} & S_1 & S_2 & S_3 \end{array} \\ \left[\begin{array}{cccc} S_1 & P_{11} & P_{12} & P_{13} \\ S_2 & P_{21} & P_{22} & P_{23} \\ S_3 & P_{31} & P_{32} & P_{33} \end{array}\right] \end{array}$$

Figure 11. Markov transition matrix

Present state		*Future state* S_1	S_2	S_3
Post-fire vegetation	S_1	0	1	0
Scrub vegetation	S_2	0	0.8	0.2
Woodland vegetation	S_3	0.1	0	0.9

Figure 12. Hypothetical transition matrix for vegetation succession

As an example, consider the hypothetical transition matrix of Figure 12, where the transitions are assumed to take place over a long period of time, say 20 years. Only three states are included in this example, so that 3×3 transition matrix is sufficient to describe the process completely, where S_1 represents post-fire vegetation, S_2 represents scrub vegetation, and S_3 represents woodland vegetation. This matrix defines the only possible transition from post-fire vegetation as one to scrub vegetation. There is absolute certainty ($P_{12} = 1$), therefore, that post-fire vegetation will be succeeded by scrub vegetation ($P_{13} = 0$). Where scrub vegetation already exists, however, there is a zero probability of a transition to post-fire vegetation, a probability of 0.8 of remaining as scrub, and a probability of 0.2 of a transition to woodland. When a woodland vegetation has been established, there is a probability of 0.9 of the woodland vegetation remaining, and a probability of 0.1 of returning to post-fire vegetation, but a zero probability of returning to scrub. The system can only reach the scrub vegetation state after passing through the post-fire vegetation state.

An alternative method of representing the transition probabilities of a Markov matrix is the Markov transition diagram shown in Figure 13. The arrows from each state indicate the possible states to which a transition may be made, and the values beside each arrow show the probability of the transition being made.

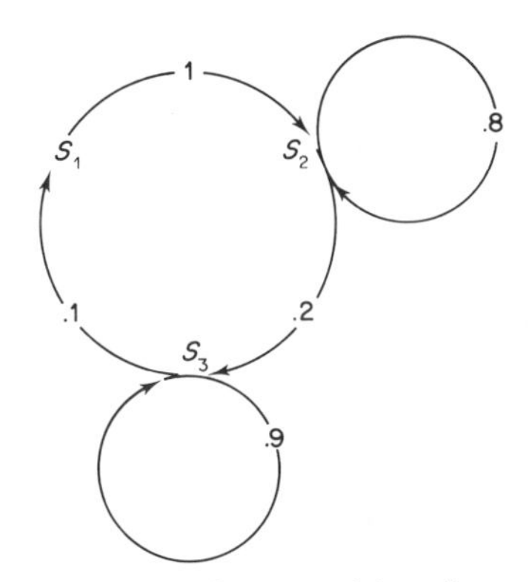

Figure 13. Markov transition diagram

It is not argued that the transition probabilities in this example are representative of any particular ecosystem. For any real application, the probabilities would have to be based on observed frequencies, derived from long-term monitoring of changes taking place in post-fire vegetation. It can be shown that some actual sequences in vegetation succession exhibit the Markov property. Where this property can be demonstrated, the transition matrix is a useful analytical tool, providing a simple way of describing in probabilistic terms a succession of events in time or space. An alternative use of the transition matrix, however, is as a regulating mechanism in the construction of simulation models of the dynamics of change in ecosystems, and examples of this type of application are given below.

6.3 CALCULATION OF TRANSITION PROBABILITIES

The construction of Markov models requires three broad groups of information:

(i) some classification that, to a reasonable degree, separates successional states in time or space into definable categories (the multivariate models of the next chapter are frequently useful in establishing such states);
(ii) data to determine the transfer probabilities or rates at which states change from one category of this classification to another with time;
(iii) data describing the initial conditions at some particular time, usually following a well-documented perturbation.

Transition probabilities are commonly based on frequency distributions or tabulations of the number of transitions from each state to each other state in the system under consideration. The frequencies are converted to estimates of the probabilities by dividing each row by its total. Alternatively, the transition probabilities may be computed from a series of integers representing different states by the simple BASIC algorithm given in the Appendix. Input to the program consists of a sequence of integers in which the value of each integer denotes a state. Thus, if there are six states in the sequence, the integers 1 to 6 might be used.

A test for the Markov property has been suggested by Anderson and Goodman (1957) as:

$$-2 \log_e \lambda = 2 \sum_{i,j}^{m} n_{ij} \log_e[P_{ij}/P_j]$$

where P_{ij} = probability in cell i, j of the transition probability matrix

P_j = marginal probabilities for the jth column

$$= \sum_{i}^{m} n_{ij} \Big/ \sum_{i,j}^{m} n_{ij}$$

n_{ij} = transition frequency total in cell i, j of the original count of the observed transitions

m = total number of states

The test distinguishes between the two alternative hypotheses:

H_0 : the successive events are independent of each other
H_1 : the successive events are not independent

If the null hypothesis (H_0) is rejected, the successive events may form a first-order Markov chain. The value $-2 \log_e \lambda$ is distributed asymptotically as χ^2 with $(m-1)^2$ degrees of freedom. A BASIC algorithm for this test is given in the Appendix.

It is convenient to assume that Markov transiton matrices are the result of processes that are stationary in time or space, i.e. that the transition probabilities do not vary with either time or space. If a very long sequence of observations is available, or if several sequences of observations from different locations are available, separate transition matrices can be calculated for each sub-interval or location. For a stationary process, these matrices should at least be similar, and Anderson and Goodman (1957) also suggested a test for stationarity. In a stationary Markov chain, P_{ij} is the probability of a transition from state i at time $t-1$ to state j at time t. In a non-stationary Markov chain, the transition probabilities vary with time (or space). Thus, $P_{ij}(t)$ is the probability of a transition from state i to j, and is a function of time (or space). The null hypothesis to be tested is that $P_{ij}(t) = P_{ij}$ for all $t = 1, 2, \ldots, T$. In other words, the test is to verify that the transition probabilities calculated from each sub-interval of time (or space) are equal to the pooled transition probability matrix obtained by estimation over the whole sequence. The alternative hypothesis is that the Markov process is non-stationary, i.e. $P_{ij}(t) \neq P_{ij}$.

The suggested test is:

$$-2 \log_e \lambda = 2 \sum_{t}^{T} \sum_{i,j}^{m} n_{ij}(t) \log_e [P_{ij}(t)/P_{ij}]$$

where m = number of states
T = number of sub-intervals
$n_{ij}(t)$ = frequency count for the transition from state i to state j in the tth sub-interval.

The value $-2 \log_e \lambda$ is distributed as χ^2 with $(T-1)(m(m-1))$ degrees of freedom. If the null hypothesis of stationarity is not to be rejected, the calculated value of χ^2 must be less than the tabulated value at some pre-selected level of significance for the appropriate number of degrees of freedom. A BASIC algorithm for the test is given in the Appendix.

$$P^{(n)} = \begin{bmatrix} P_{11}^{(n)} & P_{12}^{(n)} & P_{13}^{(n)} \\ P_{21}^{(n)} & P_{22}^{(n)} & P_{23}^{(n)} \\ P_{31}^{(n)} & P_{32}^{(n)} & P_{33}^{(n)} \end{bmatrix}$$

Figure 14. Three-state Markov chain with n time steps

6.4 POWERS OF MARKOV TRANSITION MATRICES

Although we have so far only considered a single step in a Markov chain, the probabilities of multiple step transitions can be calculated readily by multiplying the transition matrix by itself an appropriate number of times. If the system begins in state P_i, the probability that it will be in state P_j after n steps may be denoted by $P_{ij}^{(n)}$. Note that this notation does not indicate the nth power of the element P_{ij}, but indicates instead the probability of passing from state i to state j in n time steps. Figure 14, therefore, represents a three-state Markov chain after n steps.

It can easily be shown that the values of the probabilities for a transition matrix after two successive steps are given exactly by multiplying the one-step transition matrix by itself, or $P^{(2)} = PP$. Similarly, a three-step transition may be written as $P^{(3)} = P^{(2)}P$, and, in general, for the nth step, we may write $P^{(n)} = P^{(n-1)}P$. An algorithm for the successive powering of transition matrices is given in the Appendix.

6.5 CLASSIFICATION OF STATES AND MARKOV CHAINS

Before exploring the kinds of analysis made possible by the use of Markov chains it is important to identify the components of a simple classification of Markov chains. In particular, it is necessary to distinguish between two principal kinds of states, transient and closed. A composite Markov chain may be composed of several transient states and one or more closed states.

A *closed* set contains one or more states that have the property of confining a Markov process once it enters any of the states of the set. If the closed set contains more than one state in which communication between states is possible, the Markov process can move from state to state within the set, but it can never leave the set. If more than one closed state is present, the Markov process will eventually be confined within one of the closed sets, and will now not be able to reach any other closed sets. If only one state is present in a closed set, that state is an *absorbing* state, and the Markov chain is an absorbing Markov chain.

A *transient* set contains only states that are temporary and, where combined with a closed set, the transient set leads the process towards the closed set. However, not all Markov chains have a transient set or sets. A matrix containing more than one closed set, but no transient set, necessarily consists

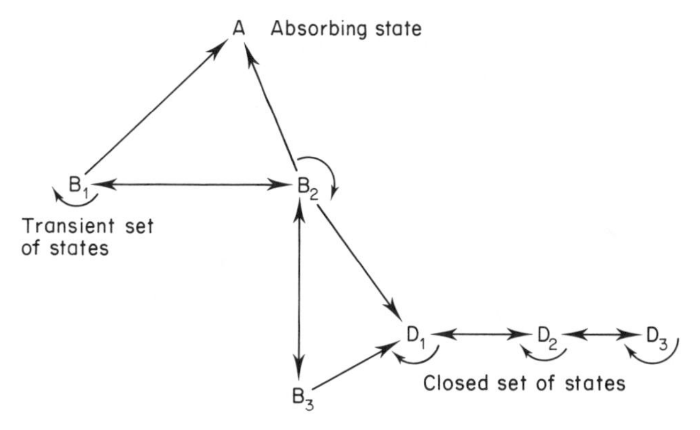

Figure 15. Diagrammatic representation of transient, closed and absorbed states

of two or more unrelated Markov chains that have been considered together needlessly. Without the transient set, there can be no communication between the closed sets, and the chains represented by the different sets can be studied separately. A Markov process without an absorbing state represents a process that is constantly in transition, and is known as an *ergodic* Markov chain.

Figure 15 gives a diagrammatic representation of transient, closed and absorbing states. State A is an absorbing state. Once the process enters this state, it does not leave it. Similarly, the states D_1, D_2, and D_3 represent a closed set. Having entered D_1, the process can move to D_2 or D_3, but cannot make a transition to any other state. In contrast, the states B_1, B_2 and B_3 represent a transient set, linking the absorbing state A to the closed set D.

By rearranging the order of the states, or by algebraic analysis, it is always possible to partition a transition matrix so as to demonstrate the existence of transient sets of states, closed sets of states, or absorbing states, as in

$$\mathbf{P} = \begin{array}{c} \begin{array}{cc} r-s & s \end{array} \\ \left[\begin{array}{c|c} \mathbf{S} & \mathbf{0} \\ \hline \mathbf{R} & \mathbf{Q} \end{array}\right] \end{array} \begin{array}{l} \\ r-s \\ s \end{array}$$

where s = number of transient states
$r - s$ = number of closed states
Q = square sub-matrix, containing $s * s$ elements, for transitions between transient states
R = rectangular sub-matrix containing $s * (r - s)$ elements, for transitions from transient set to closed set or sets
S = square sub-matrix containing $(r - s) * (r - s)$ elements, forming a class of closed sets from which there is no exit after entry
0 = sub-matrix, containing $s * (r - s)$ elements, all of which are zero and therefore represent no transition

Figure 16. Analysis of transition matrices

Figure 16. Where a transition matrix can be partitioned in this way, the several components can be investigated separately, thus simplifying the task of studying and modelling the ecosystem. Analysis of the matrix of transition probabilities also simplifies the calculation of the average times needed to move from one state to another, and of the average length of stay in a particular state once it has been entered. Where closed or absorbing states exist, the probability of absorption and the average time to absorption can also be calculated readily. Examples of these calculations are shown for a simple model below.

6.6 ANALYSIS OF AN ERGODIC MARKOV CHAIN

Raised mires frequently show interesting successional changes as a result of increased drainage, and Figure 17 gives the estimated probabilities for the transitions between four possible states of a raised mire over a period of 20 years. State 1 represents the wettest facies dominated by *Sphagnum*, with *Calluna*, *Erica tetralix* and *Eriophorum* as the major vascular plant components. State 2 represents a drier facies, with a *Calluna–Cladonia* association and seedlings of *Betula* and *Pinus sylvestris*, the more mature woodland of State 3 having a typical *Vaccinium myrtillus* community with hypnaceous mosses. State 4 represents disturbance due to grazing by large herbivores of the drier facies, leading to the establishment of a *Molinia–Pteridium*-dominated association.

Thus, areas which start as typical bog vegetation have a probability of 0.65 of remaining as bog vegetation at the end of the 20-year time-step, and probabilities of 0.29 and 0.06 respectively of becoming *Calluna*-dominated and woodland. Areas which start as *Calluna*-dominated have roughly equal probabilities of remaining in the same state, returning to bog vegetation because of fluctuations in the water table, or of becoming woodland: they have a small (0.07) probability of being subjected to sporadic grazing. Woodland areas have a 0.69 probability of remaining as woodland, a probability of 0.28 of returning to *Calluna* because of deaths of trees, and, again, a small (0.03)

Starting state	*Probability of transition to:*			
	1. Bog	2. *Calluna*	3. Woodland	4. Grazed
1. Bog	0.65	0.29	0.06	0.00
2. *Calluna*	0.30	0.33	0.30	0.07
3. Woodland	0.00	0.28	0.69	0.03
4. Grazed	0.00	0.40	0.20	0.40

Figure 17. Transitional probabilities for successional changes in a raised mire (time step = 20 years)

probability of being subjected to sporadic grazing. The grazed areas have an equal probability of being subjected to further grazing and of returning to a *Calluna*-dominated vegetation, and a small (0.20) probability of becoming woodland because of the growth of ungrazed seedlings.

None of the states, therefore, are absorbing or members of a closed set, and the matrix represents a transition from the bog vegetation to woodland, with an imposed disturbance due to grazing. However, although there can be a return from the *Calluna*-dominated vegetation to bog vegetation because of fluctuations in the water table, there is no immediate return from woodland to bog vegetation. Where there are no absorbing states, the Markov process is known as an ergodic Markov chain, and the full implications of this model can be exploited by an analysis of the transition matrix.

Using the algorithm of the Appendix, it is possible to calculate the transition probabilities after two, three, four, ..., time steps. Thus, for the transition matrix of Figure 17, the corresponding probabilities after two time steps are:

$$\begin{bmatrix} 0.5095 & 0.3010 & 0.1674 & 0.0221 \\ 0.2940 & 0.3079 & 0.3380 & 0.0601 \\ 0.0840 & 0.2976 & 0.5661 & 0.0523 \\ 0.1200 & 0.3480 & 0.3380 & 0.1940 \end{bmatrix}$$

and after four time steps are:

$$\begin{bmatrix} 0.3648 & 0.3035 & 0.2893 & 0.0424 \\ 0.2759 & 0.3048 & 0.3649 & 0.0543 \\ 0.1841 & 0.3056 & 0.4528 & 0.0595 \\ 0.2151 & 0.3114 & 0.3946 & 0.0789 \end{bmatrix}$$

Alternatively, the Appendix gives a BASIC program which will calculate directly the transition probabilities after a given number of time steps.

If a matrix of transition probabilities is powered successively until a state is reached at which each row of the matrix is the same as every other row, forming a fixed probability vector, the resulting matrix is called a *regular* transition matrix. The matrix then gives the limit at which the probabilities of passing from one state to another are independent of the starting state, and the fixed probability vector expresses the equilibrium proportions of the various states. In the example above, the vector of probabilities is:

$$[0.2177 \quad 0.2539 \quad 0.3822 \quad 0.1462]$$

If, therefore, the transition probabilities have been correctly estimated and remain stationary, the raised mire will eventually reach a state of equilibrium in which approximately 22% of the mire is bog, and approximately 25%, 38% and 15% are *Calluna*, woodland and grazed communities, respectively.

Although the limiting or equilibrium probabilities can be calculated by successive powering of the transition matrix, a quicker method is given in the

$$\begin{bmatrix} 0 & 3.561 & 7.197 & 31.688 \\ 9.566 & 0 & 5.237 & 28.755 \\ 13.672 & 4.107 & 0 & 29.178 \\ 18.673 & 9.107 & 5.000 & 0 \end{bmatrix}$$

Figure 18. Mean first passage times for raised mire system

BASIC algorithm of the Appendix. The same program calculates the mean first passage times defined as the average length of time required to move from any one state to one of the other states. Figure 18 gives the mean first passage times for the transition matrix of Figure 17. Each element of the mean passage time matrix needs to be multiplied by the time step of 20 years. Thus, the average length of time a *Calluna*-dominated area takes to become bog is $9.566*20 = 191$ years. Similarly, the average length of time needed for woodland to become *Calluna* is $4.107*20 = 82$ years, and the other times can be calculated as required.

Alternatively, if an area is chosen at random, the average length of time needed for that area to reach any of the defined states is given by the mean first passage times in equilibrium, also calculated by the program of the Appendix. For the raised bog system, the mean first passage times in equilibrium are given by the vector:

$$[10.385 \quad 3.676 \quad 3.627 \quad 25.351]$$

Remembering again that each time step represents 20 years, the mean first passage time for a randomly chosen area to become bog is $10.385*20 = 208$ years, while the corresponding mean first passage times for *Calluna*, woodland and grazed communities are 74, 73 and 507 years, respectively.

6.7 ANALYSIS OF AN ABSORBING MARKOV CHAIN

In the analysis of an absorbing Markov chain, in contrast to that of an ergodic chain, the main emphasis is necessarily focused on the average time for the

Starting state	*Probability of transition to:* Woodland	Bog	*Calluna*	Grazed
Woodland	1.00	0.00	0.00	0.00
Bog	0.06	0.65	0.29	0.00
Calluna	0.30	0.30	0.33	0.07
Grazed	0.20	0.00	0.40	0.40

Figure 19. Modified transition probabilities for successional changes in a raised mire (time step = 20 years)

$$\begin{bmatrix} 4.75 & 2.21 & 0.26 \\ 2.88 & 2.67 & 0.31 \\ 1.52 & 1.78 & 1.87 \end{bmatrix}$$

Figure 20. Fundamental matrix for raised mire transition matrix modified to create an absorbing state

system to reach the absorbing state, and on the probability that the system will reach this state from any given starting state. Figure 19 gives a version of the transition probabilities for the raised mire system, modified so that the woodland state becomes an absorbing state. It is then convenient to rearrange the matrix so that the absorbing state is partitioned from the set of transition states.

The appropriate analysis for an absorbing Markov chain is given by the algorithm of the Appendix. This algorithm first calculates the so-called *fundamental* matrix which summarizes the number of time steps for which the system will be in the state indicated by the column before being absorbed, given that the present state is that of the row. The fundamental matrix for the modified raised mire system is given in Figure 20 and indicates, for example, that 2.21 is the number of time steps that the system will, on average, be in the *Calluna* state, having started as bog, before being absorbed into woodland. Where there is only one absorbing state, the probability of reaching that state is necessarily equal to 1.0, but, where there are two or more absorbing states, it is important to determine the probability of reaching each of these states from each of the possible starting states, and the program also calculates these probabilities. Finally, the same program calculates the number of time steps, on average, that the system will take to reach the absorbing state from each of the possible starting states. For the modified transition matrix of Figure 19, these numbers of time steps are summarized in the following vector:

$$[7.22 \quad 5.27 \quad 5.18]$$

6.8 DEPENDENCE, ORDER AND STEP LENGTH IN MARKOV CHAINS

It is customary to classify Markov chains in terms of their dependence, order and step length. The Markov chains illustrated so far in this chapter have been dependent only on the immediately preceding state, and are therefore defined as being single-dependence chains. They are also first-order chains because this preceding state is immediately preceding. The transition involves a single step of unit length, equivalent to the length of the time step.

It is, however, possible to define Markov chains with dependency relationships that involve more than one preceding step, and a double-dependence chain, for example, is one that is dependent on two preceding states. If these two states are the two immediately preceding states, the chain is a second-order chain, and either or both of these steps may be of greater than unit length. An algorithm for analysing and testing for the double-dependence Markov property is given by Harbaugh and Bonham-Carter (1970). Input to the algorithm includes a sequence of states for which a transition-frequency matrix is calculated and subsequently transformed into a transition probability matrix. In addition, the algorithm calculates a maximum likelihood criterion test statistic which is similar to that for first-order Markov chains. The null hypothesis under test is that the chain is singly dependent against the alternative that it is doubly dependent. The program calculates this test statistic for each value of the second memory step length.

6.9 DISCRETE-STATE CONTINUOUS TIME MODELS

In the usual discrete-time Markov chain, the system advances in a series of discrete time steps, with transitions occurring at each of the steps. In a continuous-time Markov model, a matrix of transition rates q_{ij} is employed. These transition rates not only indicate the probability of one state succeeding another, but also establish the expected waiting time in each state.

In the continuous-time model, attention is focused less on the probability of a state change, p_{ij} (where $i = j$), than on the rate of transfer, q_{ij}, from state i to state j. The transition rates may be obtained by direct observation, or alternatively they can be derived by algebraic transformation of the p_{ij} matrix. The following equations relate transition probabilities to transition rates:

$$p_{ij} = e^{-M\Delta t}$$

$$p_{ij} = \frac{q_{ij}}{M_i} p_i \qquad (i \neq j)$$

$$q_{ij} = \frac{p_{ij}}{p_i} M_i \qquad (i \neq j)$$

where M_i = sum of the off-diagonal elements in the ith row of the q_{ij} matrix

$$\sum_{\substack{j=1 \\ j \neq i}}^{k} q_{ij}$$

and k is the number of states.

p_i = sum of the off-diagonal elements in the ith row of the p_{ij} matrix

$$\sum_{\substack{j=1 \\ j \neq i}}^{k} p_{ij} \qquad \text{or } [1 - p_{ij}]$$

where p_{ij} is the diagonal element.

Rearranging the first of these equations:

$$M_i = \frac{-\log_e p_{ij}}{\Delta t}$$

The reverse transformation from the q_{ij} matrix to the p_{ij} matrix can be made by using the same equations. In this case, the values of q_{ij} are known, and the values of M_i are calculated by summing each row of the q_{ij} matrix. The values of p_{ij} are then determined directly from the equations. Harbaugh and Bonham-Carter (1970), from whom this section is taken, also give a FORTRAN algorithm for accomplishing these transformations in either direction.

6.10 GENERATING SEQUENCES OF EVENTS FROM MARKOV CHAINS

In addition to the analysis of ergodic and absorbing Markov chains described above, it is often of interest to generate sequences of events from a matrix of transition probabilities or rates. By their very nature, involving probabilities, the outcome of such sequences will vary from simulation to simulation, and this variation may be exploited to show the essential variability of the system, possibly derived from genetic or environmental causes.

The simplest of these sequence generation procedures is that depending on a simple discrete event, discrete-time transition matrix. At each time step, the program refers to the given transition matrix and determines the state to which the system moves by generating a random number. This computation can be repeated as often as is necessary to determine what may be regarded as the 'final' state. Where a two-dimensional system is being simulated, the result may be a computer-generated map.

The principles of simulating double dependence Markov chains are identical to those used in generating single dependence chains. Input to the program includes the number of states, the length of the second step in whole-number multiples of the unit time step, the number of transitions to be generated, and the transition probabilities. Several two-dimensional double dependence transition probability matrices may be read and stored in a multi-dimensional array.

6.11 MARKOV-TYPE MODELS, BASED ON DIFFERENTIAL EQUATIONS

As an alternative to the use of conventional Markov chains, it may be convenient to employ the commonly used and well-documented techniques of feedback and control systems, by using ordinary differential equations to describe the trajectories leading from one state to another. Shugart *et al.* (1973) give an example of this kind of approach in the modelling of forest succession over large regions. With this procedure, it is necessary to define a set of possible cover states, or stand types, in the same way as for the state-by-state approach of the Markov model.

The transition probabilities used in the Markov model and the rate constants used in a system of differential equations can be calculated from the same data set. Both types of model assume that the replacement of one type of stand by another is dependent only on the present conditions, and that replacement patterns remain constant with time and are not affected by spatial heterogeneity. The approach is therefore essentially Markovian in character, and the results from a conventional Markov model and from the system described by differential equations, also derived from the same data set, should be identical.

The choice between the two approaches may often depend on the objectives of the study, as well as on the knowledge and mathematical expertise of the user, but, when a straightforward Markov approach can be used, it is usually to be preferred. Not only do the possibilities of further algebraic analysis lead to a better appreciation of the stochastic nature of many ecological processes, but, by the calculation of such parameters as the mean passage time, time to absorption, and the degree of stability and convergence within the defined states, additional information of direct ecological and management value is provided.

Extensions of the Markov model suggest the use of higher-order chains, where the next state is dependent not only on the present state, but also on one or more previous states. Processes may also be regarded as semi-Markovian, the time spent in a given state (the direct passage time) being variable and dependent on the present and the next state. Certainly, some biolgoical processes might be more accurately represented in this way. Although these modifications of the Markov model illustrate some of the flexibility that is conceptually possible within a basically Markovian approach, the collection of data and the development of appropriate transition matrices become much more difficult.

Data collection for the calculation of transition probabilities and for the construction of transition matrices is a major task, ideally requiring detailed documentation of changes in systems over extended periods of time, and of responses to various types of perturbation. Nevetheless, data of this type do exist, both from historical and experimental records. In other situations, it may be sufficient to use hypothetical transition probabilities so as to show the consequences of particular rates of transition for the future development of a system.

In contrast, when a differential equation approach is used, the analytical techniques developed to explore the control and feedback aspects of such models are directly applicable, and themselves provide useful insights into ecosystem management.

It is, perhaps, worth emphasizing the relationship between the Markov models described in this chapter and the Leslie matrix models described in Chapter 5. The Leslie matrix describes the transition probabilities from one time to another, the state of the population at time $t + 1$ being dependent on the state of the population at time t. However, the probabilities in the Leslie matrix do not sum to unity in either the rows or the columns, and the Markov theory is not therefore applicable to these matrices, despite their apparent similarity.

6.12 ADVANTAGES AND DISADVANTAGES OF MARKOV-TYPE MODELS

The advantages of Markov-type models may be summarized as follows.

(i) Markov models are relatively easy to derive (or infer) from successional data.
(ii) The Markov model does not require deep insight into the mechanisms of dynamic change, but it can help to indicate areas where such insight would be valuable and hence act as both a guide and stimulant to further research.
(iii) The basic transition matrix summarizes the essential parameters of dynamic change in a way that is achieved by very few other types of model.
(iv) The results of the analysis of Markov models are readily adaptable to graphical presentation, and, in this form, are frequently more readily presented to, and understood by, resource managers and decision-makers.
(v) The computational requirements of Markov models are modest, and can easily be met by small computers, or, for small numbers of states, by simple calculators.

The disadvantages of Markov models are as follows.

(i) The lack of dependence on functional mechanisms reduces their appeal to the functionally orientated ecologist.
(ii) Departure from the simple assumptions of stationary, first-order Markov chains while, conceptually possible, makes for disproportionate degrees of difficulty in analysis and computation.
(iii) In some areas, the data available will be insufficient to estimate reliable probability or transfer rates, especially for rare transitions. For example, it may not be possible to observe sufficient transitions from a given transient set of states to a closed state where this transistion is dependent on a rare climatic event, even though the value of this parameter is of vital importance in the dynamics of the community.
(iv) Like all other successional models, the validation of Markov models

depends on predictions of system behaviour over time, and is therefore frequently difficult, and may even be impossible, for really long periods of time.

These difficulties warrant a high degree of caution in the uncritical use of Markov models in ecology. Of the disadvantages listed, the second is perhaps of the greatest significance because it is doubtful whether most ecological successions do, in fact, have homogeneous transition matrices that are constant in time. A constant transition probability implies that the future behaviour of a system is determined only by its present state and is independent of the way in which this state has developed. It also implies that the replacement of, for example, an individual of species j by one of species k does not vary over the time units used in calculating the transition probabilities.

Both of these assumptions run counter to many widely accepted biological ideas concerning evolution, adaptive strategies and the occurrence of episodic events such as abnormal weather, natural catastrophes and epidemics. However, the accumulating evidence that the initial floristic composition, after a perturbation, dominates the subsequent successional patterns supports the view that past history may not be of primary importance in determining successional sequences following the perturbation, even though it is of a major factor in influencing the suites of species that exist at a site and are therefore available to participate in the succession. Horn (1975) has suggested that the episodic event, shifting the course succession from one pathway to another, may well be included in the Markov model by switching from one transition matrix to another.

6.13 CASE STUDIES OF MARKOV MODELS

Although the mathematical basis of the Markov model has been well known to mathematicians for many years, there have been relatively few practical applications of the model to studies of ecosystem dynamics. Whether this lack of application is due to the inaccessibility of the theory to non-mathematicians or to the practical difficulties of collecting data from which to estimate the transition probabilities is not clear, and perhaps both of these reasons have contributed to the slowness of practical use of the technique in ecology.

One of the early applications of Markov models to ecology was by Williams *et al.* (1969) in their study of rainforest communities. In these studies, ten sites were cleared in tropical forest, and the presence and absence of species on twelve occasions over a period of approximately seven years were recorded, the records being subsequently classified into seven states. By counting the frequency with which one state followed another, the transition probabilities of a Markov process were estimated. Stephens and Waggoner (1970) and

Waggoner and Stephens (1970) used a similar approach to predict the composition of a mixed hardwood forest in the eastern United States from the enumeration of natural transitions over a 40-year period. Some of the best-known examples, however, of the use of Markov models to characterize forest succession were given by Horn (1974, 1975), in which the succession of species was estimated by simple approximations of the tree-by-tree replacement within the stand. Extrapolation of the Markov model gave a forest composition which was very similar to areas of forest which were known to be very old, and which were believed to be largely untouched by man.

Other studies of the use of Markov models in forests include those of Peden *et al.* (1973), Cassell and Moser (1974), and Lembersky and Johnson (1975). More recently, Binkley (1980) has applied stationary Markov models to three kinds of problems arising in forestry, including individual tree level models of forest succession, plot/canopy gap level models of forest succession, and stand level analyses of specific forest management problems. Buongiorno and Michie (1980) have also developed a Markov model of a selection forest, where the parameters of the model represent the stochastic transition of trees between diameter classes and the recruitment of new trees into the stand. The parameters of this model were estimated from north-central United States region hardwoods, and the model was used to predict long-term growth of undisturbed and managed stands. Subsequently, a linear programming method was used to determine sustained use of management regimes which would maximize the net present value of periodic harvests. The method allows for the joint determination of optimum harvests, residual stock, diameter distribution and cutting cycles.

Applications to other types of ecological system are even rarer than those of forests, but Debussche *et al.* (1977) describe an agricultural application to the southern end of the French Massif Central. This model considered the rural exodus and decrease of sheep grazing, and the best utilization of the grazing resources. The mapping of vegetation types, study of the dynamic changes from one type to another and the definition of key species were used as a basis for the formulation of the model. Vandeveer and Drummond (1978) have also used Markov processes for estimating land use changes, particularly where a major impact such as a reservoir is imposed upon an existing system.

Other biological applications of Markov processes include that by Rao and Kshirsagar (1978), who made a study of the population dynamics of predator/prey systems in which the attack cycle of a predator is assumed to consist of four different activities, namely search, pursuit, handle and eat, and digestion. A semi-Markovian model was proposed to obtain the number of prey devoured by a predator during the activity of a day. Usher and Parr (1977) suggest that two kinds of succession could be recognized in a decomposer communities, one related to habitat change (plant species, microenvironment), and the other to the decomposition stages of the decaying food resource. Data are given for the decomposition of wood by termites in West Africa, and a succession of both plants and soil arthropods on the developing chalk soil in Britain, and it is suggested that both processes are Markovian in character. Usher (1979), in an analysis of published studies, suggests that ecological succession can usually be considered as a non-random process, and that complex non-random or Markovian processes are likely to characterize all ecological successions. The elements of the transition probability matrix are, however, unlikely to be constant, but are functions of either the abundance, or the rate of change in abundance, of a recipient class. Glaz (1979) provides explicit formulae for the absorption probabilities, means and variance of first absorption times in terms of birth and death rates in finite homogeneous birth and death processes.

Tavare (1979) has proposed a simple and intuitive method of deriving some properties of finite homogeneous continuous-time Markov chains in population genetics. The chains have absorbing barriers, and the method involves the representation of such a process in terms of a discrete Markov chain, and a series of waiting times which reconstruct the original timescale.

Although not strictly applied to ecological problems, the text by Bartholomew and Forbes (1979) gives practical applications of Markov and Markov-type methods in manpower planning, and many of the techniques which are suggested are of practical value in the modelling of biological populations. The general-purpose programme included for showing the development of a population from a given starting point is of particular value. Finally, Cook (in press) gives an interim report on an attempt to estimate the transition probabilities of plant-by-plant replacement from aggregate data, i.e. the percentage of sites in each state at regular time intervals, and a least-squares method is proposed for estimating these transition probabilities, but the method is not, as yet, entirely successful.

Geologists have used Markov models to provide a mechanism for computer simulation of a wide variety of geological processes, and Krumbein (1967) emphasizes the value of first-order Markov chains because of their intuitive appeal, particularly in stratigraphic analysis. The text by Harbaugh and Bonham-Carter (1970) contains a chapter giving a general account of Markov models, with applications to geology and stratigraphy. This chapter contains several useful FORTRAN programs for critical computations. Norris (1971) gives a comparison of different methods in the transition matrix approach to the numerical classification of soil profiles. Bhattacharya *et al.* (1976) suggests a Markovian stochastic basis for the transport of water through unsaturated soil, and Lloyd (1977) provides a theoretical proposal for the modelling of reservoirs by a first-order Markov process. No actual example is given of the application of these processes, however, and the major part of the paper is devoted to the problem of obtaining the probability distributions of recurrence times and first passage times. Smeach and Jernigan (1977) present further aspects of a Markovian model sampling policy for water quality monitoring.

Finally, Anthony and Taylor (1977) explore the use of Markov models in forecasting air pollution levels. The analysis of historical data concerning variations in air pollution indices suggests a pattern which might usefully be described by a transition probability matrix. By utilizing what is essentially a Markov-based framework, the model suggests a number of interesting and important questions and facilitates their solutions. In their presentation, the authors focus primarily on the variations in a single component of air pollution, i.e. suspended particulate matter. For implementation of the model it would be necessary to handle other components in the same manner.

CHAPTER 7

Multivariate Models

Strictly speaking, multivariate models are not uniquely dynamic. However, they enable the ecologist to explore several, and often many, dimensions simultaneously, and one of these dimensions may well be time. Alternatively, the several dimensions modelled in a multivariate analysis may be related to time as an explanatory variable, and so provide an explanation, and in some cases prediction, of the changes that take place with time. In this chapter, a review of multivariate models will be given, drawing on the very extensive literature that now exists on this topic. Some hints are given about the computational aspects of multivariate models, but the reader will often be referred to texts where very extensive advice is found. Similarly, the theory of multivariate models is not discussed in detail in this chapter, because there are a great many texts where this theory can be found to any desired level of detail and mathematical complexity.

7.1 ORIGINS AND DEVELOPMENT OF MULTIVARIATE MODELS

Much of the mathematics involved in the construction of multivariate models is not new. For example, the fundamental probability distributions related to the multivariate normal distribution were derived in the 1930s, and the methods developed then are the basis for most of the multivariate methods used today. However, the calculations involved in multivariate analysis, and in the construction of multivariate models, becomes very tedious when the number of variates is large, and more of these computations were virtually impossible to carry out for more than four or five variates, even on electrically driven calculating machines. Until the electronic computer became generally available, therefore, only a very limited number of extensive multivariate analysis has been attempted, and the same examples were quoted in amost all of the earlier textbooks.

The development of computers has completely changed the situation, with the result that multivariate models have now become an important addition to the range of models which may be used by the ecologist. Many of the computations can be done on quite simple microprocessors, so that the application of multivariate models is not restricted to ecologists with access to large mainframe computers. Indeed, there can often be distinct advantages in modelling multivariate data in several stages, with careful examination of the results at each stage, before proceeding to the next stage of the analysis. In consequence, a rapidly increasing collection of examples has appeared in the scientific literature, almost too many to review in one chapter of this Handbook. Nevertheless, these multivariate methods represent a much neglected class of models, especially in dynamic or time related applications.

Many of the multivariate techniques available today have been developed in association with taxonomy and other branches of the more descriptive sciences. This association has had several interesting and unexpected effects. For example, what used to be largely intuitive has become increasingly more formalized and more quantitative. The sudden explosion of computers and computer science has enabled scientists to explore a wide range of techniques, and this new field of activity has attracted the attention of statisticians and mathematicians, with the consequent development of an even wider variety of methods and of applications to problems in many different scientific fields. This rapid development has not been an unmixed blessing, as there now exists a bewildering variety of numerical techniques, the properties of many of which are not adequately known. Furthermore, while numerical methods can often greatly help the ecologist to investigate the structure of his data, the results of the analyses provided by these methods still need to be interpreted, and there are still very few ecologists with experience in such interpretation.

It is relatively easy to perceive the structure of multivariate data when discontinuities are obvious, but such a situation is not typical of many ecological applications. Much of what we observe in nature changes more or less continuously in one or more properties, but not necessarily by the same degree in each of the properties. It is these more gradual changes which provide the greatest problems for the scientist to decide where or how to draw boundaries, or, indeed, whether to draw any boundaries at all. The aim of this chapter, therefore, is to clarify some of the issues involved in the use of numerical methods to search for structure and dynamic change in ecological data. Ecological applications are less well developed than those in plant and animal taxonomy, partly because of the diversity of interests of ecologists, and partly because of the nature of the ecological data themselves. The methods appropriate to the taxonomy of organisms are not, therefore, necessarily the most appropriate to dynamic ecological problems (Clifford and Stephenson, 1975).

By virtue of the speed with which they can process large quantities of data and perform complex calculations, computers have revolutionized the handling of scientific data. However, there is a danger that the power of computers sometimes leads those unfamiliar with the mathematical theory of the methods to assume that what is produced by the computer necessarily contains some objective 'truth' which should not be questioned. Nowhere is this more the case than in multivariate analysis. Anderberg (1973) has commented that

> cluster analysis methods involve a mixture of imposing a structure on the data and revealing the structure which actually exists in the data. The notation of finding natural groups tends to imply that the algorithm should passively conform like a wet tee-shirt. Unfortunately, practical procedures involve fixed sequences of operations which systematically ignore some aspects of the structure while intensively dwelling on others.

The use of a computer does not release the ecologist from thinking; quite the reverse. Multivariate models, in particular, lay traps for the unwary, and it is the purpose of this chapter to spring many of these traps.

7.2 ATTRIBUTES, VARIABLES AND VARIATES

Multivariate data in ecology, as in other applied sciences, consist of sets of attributes or scores for each of a number of variables, this number being greater than two and sometimes large. Conventionally, the term *attribute* is used by statisticians to represent the qualities possessed or not possessed by an individual. The term *variable* is used for qualities which are measured quantitatively along some continuous scale. The strict definition is of a quantity that may take any one of a specified set of values, where these values may be continuous, as in measurements of height or weight, or discontinuous, as in counts of individuals. By extension, the term is sometimes also used to denote non-measurable characteristics. For example, *sex* may be regarded as a variable in a sense, as any indiviual may take one of two values — male or female.

A *variate* is a quantity which may take any one of the values of a specified relative frequency or probability. Such variates are also known as random variables and they are regarded as being defined not merely by a set of permissible values like any ordinary mathematical function, but by an associated frequency or probability function expressing how often these values appear in the application under discussion. There are many situations in ecology where models have to capture the behaviour of more than one variate, and such models are therefore known collectively as *multivariate*, an expression which is used rather loosely to denote the analysis of data which are multivariate in the sense that each individual bears the values of several variates. By extension, some of these values may be variables rather than variates.

This rather general description of attributes in ecology, as elsewhere, is convenient, as it avoids the need to differentiate between continuous and discrete data in general discussion where the nature of the data is not in question. Nevertheless, formal discussion of multivariate methods often focuses on the different kinds of attributes (see, for example, Sneath and Sokal, 1973; Clifford and Stephenson, 1975; Williams, 1976). The most common distinctions are made between: (i) binary attributes, e.g. presence or absence; (ii) disordered multi-state or nominal attributes, such as colour or soil type; (iii) order multi-state or ordinal attributes, e.g. rare, common, abundant, etc.; (iv) meristic or discontinuous, e.g. number of petals; and (v) continuous or numeric attributes. These categories are not, however, distinct, because they depend to some extent on the sampling procedures used, and data in one form can often be transformed into another.

There are marked differences of opinion about the value of binary data in ecology, but the consensus of opinion seems to be that other forms of data are preferable (e.g. Clifford and Stephenson, 1975). In most branches of ecology, some measure of dominance is often regarded as important in describing vegetation, so that the results of using analyses of numerical data regarded as being more informative than those of binary data. For example, Williams *et al.* (1973) found that while the presence or absence of plant species was adequate in a simple study involving only 8 sites, there was 'some advantage' in using the numbers of each species in a study of 10 sites, and quantitative data of the relative dominance of species were distinctly preferable for a study of 80 sites. Similarly, Barkham (1968) found quantitative data to be more informative than presence–absence in a study of a Cotswold beechwood.

While binary or presence–absence data appear to be adequate when there are major differences in species distributions between sites, they have not proved to be very useful in detailed studies of vegetation dynamics where there are relatively few species and less clear-cut differences between sites. Indeed, the use of binary data for the study of dynamic change in ecology can usually only be justified if it is difficult to obtain anything else, or if there is a declared lack of interest in the information which will be lost by the use of binary data instead of, say, continuous data. Similarly, the conversion of either meristic or continuous data to binary data is usually unsatisfactory. Division of a variable which is approximately normally distributed, for example into two parts, leads to an attribute with all the values on either side of the dividing line having identical binary scores. However, there may, in some circumstances, be instances where continuous data have properties which make conversion from one form to another logical. One such example is an attribute which can take a wide range of values, but which has a concentration at one value (usually zero), as in the counts of the number of parasites in a host species. It may then be possible to regard such an attribute as discrete, and to score it as if it were composed of a few groups, e.g. zero, low, medium, high. Nevertheless, if the aim of the analysis is to find a useful or meaningful grouping of the data, a coarsely grouped attribute may exert a disproportionate influence on the result (Marriott, 1974).

$$\mathbf{X} = \begin{bmatrix} x_{11} & x_{12} & x_{13} & \cdots & x_{1p} \\ x_{21} & x_{22} & x_{23} & \cdots & x_{2p} \\ x_{31} & x_{32} & x_{33} & \cdots & x_{3p} \\ \vdots & & & & \\ x_{n1} & x_{n2} & x_{n3} & \cdots & x_{np} \end{bmatrix}$$

Figure 21. Basic data matrix for multivariate analysis

7.3 NOTATION

Whether the basic attributes to be used in the study are binary, multi-state, discrete or continuous variables, the same notation can be employed to represent the data which will be used in any multivariate analysis. Figure 21 displays this notation as a matrix, **X**, of the values of each of p attributes for each of n individuals. Thus, x_{11} represents the value of the first attribute for the first individual, x_{32} the value of the second attribute for the third individual, and, more generally, x_{rs} the value of the sth attribute for the rth individual.

There may, of course, be *a priori* groupings or classifications of the p variables or the n individuals. Indeed, it is the presence of such additional information which may influence the choice of an appropriate multivariate technique. The existence of such information does not, however, necessitate any change in the notation.

7.4 SOME ESSENTIAL STAGES IN THE ANALYSIS OF MULTIVATE DATA

7.4.1 Choice of the attributes to be included

The first stage in any analysis is the choice of attributes to be included. It perhaps needs to be stressed that this is the most important of all the stages described. Any choice of attributes represents an hypothesis about the relevance of the variables to the solution of the problem under consideration. Much time and effort have been wasted by the application of multivariate analysis to sets of data containing a 'rag-bag' of variables included because they were easy to measure, or because they happened to be available, without any apparent consideration of the logical design of the investigation. Indeed, some scientists appear to regard multivariate analysis as a substitute for thought, expecting the methods to compensate for failures of logic which they would not permit in the analysis of univariate data (Jardine and Sibson, 1971; Sneath and Sokal, 1973; Williams, 1976).

As the selection of attributes to be included will be heavily dependent on the objectives of the research, some guidance can frequently be obtained by considering the following questions.

(i) What broad dimensions of the variability of the individuals are relevant to the problem? In seeking to describe the growth of a particular species of plant, for example, we may only be concerned with the activity of the primary meristems, and our variable will be confined to measures of shoot length, height, leaf extension, etc. We may, alternatively, be concerned with growth of both primary and secondary meristems, so that measures of diameter, sectional area or volume of various part of the plant may need to be included.

(ii) Do the variables cover the total range of the variability of the individuals in a fairly uniform manner? Inclusion of a single variable measuring vigour or health in a larger group of variables measuring size may lead to a relatively inefficient analysis.

(iii) Are some of the variables logically dependent on the others? It is not usually particularly helpful to include all of the percentages which must add up to 100, for example. Similarly, inclusion of a variable which is derived arithmetically from one or more of the others seldom adds much information to the analysis.

(iv) Is it possible to identify any structure among the variables? For example, the total set of variables may consist of two groups, one of which can be regarded as a response to the other groups in a complex investigation.

Nevertheless, having stressed the importance of the choice of the variables to be included in the analysis, there is a good deal of flexibility resulting from the use of the computers essential to the application of multivariate analysis to practical problems. Further variables may be added, if they become available or can be obtained, and variables may be added, if they become available or can be obtained, and variables may be deleted after a trial analysis, if it is found that they contribute little to the interpretation of the analysis. Indeed, one of the main advantages of computer-based multivariate analysis is that it enables the research worker to keep many variables under constant review, modifying the selection of variables as understanding of the problem increases, and eliminating those variables which contribute little to the analysis.

The number of variables that can be included in one analysis is limited only by practical difficulties of computation at critical stages of the analysis, and by the size of the computer that is available. Generally speaking, there is seldom any difficulty in dealing with up to 60 variables. For more than 60 variables, large computer systems are usually necessary, but, except for certain kinds of analysis of data consisting entirely of binary attributes, there is relatively little to be gained from including very large numbers of variables in the same analysis. It will frequently be preferable to divide large sets of variables into groups and to analyse the groups separately, investigating the relationships between the groups subsequently. Alternatively, it is sometimes possible to analyse the data by adding and discarding variables in sequential steps; such methods are not only usually more economical of computing resources, but are often more easily interpreted than the analysis of all variables simultaneously.

The number of variables is not likely to be a serious limitation in practical work. There are very few problems for which large numbers of independent dimensions of variability can be established, quite apart from the difficulty of collecting the data in the first place.

7.4.2 *A priori* weighting of attributes

The question of whether, or how, to weight data *a priori* is an important problem in taxonomy, and specialists in different groups of organisms have their own ideas about the importance of different attributes in this specific application. One solution to the problem is to state the basis of weighting, so that the reader may judge the merits of each case (Clifford and Stephenson, 1975). Sneath and Sokal (1973) consider that equal weighting is desirable; their approach can be defended on several independent grounds, and is probably the only practical solution. Jardine and Sibson (1971) conclude that certain kinds of weighting which taxonomists use intuitively are, in fact, incorporated in the calculation of many multivariate techniques as part of the analysis.

It is certainly likely that some of the attributes in any given set of data might reflect some single underlying feature, so that to use all the attributes is an implicit weighting of that feature. One way of avoiding this effect is to use principal component analysis to identify the number of independent dimensions contained in the data set, as described in Section 7.7.1.

7.4.3 Choice of the individuals to be included

Having decided on the attributes to be included in the analysis, it is necessary to consider carefully the individuals. Ideally, of course, the logic of this selection should have been decided long before any data were collected at all, as part of the design of the experiment or survey which preceded the analysis. However, as with the choice of the attributes, it frequently happens that data are presented for multivariate analysis without much thought about the individuals that have been measured or assessed.

It will commonly be assumed that the individuals are samples from some population. Only rarely will it occur that the individuals are the whole population, i.e. represent all the possible individuals for which a description is required. If the individuals are a sample from a population, has that population been defined, and can the individuals be regarded as a representative sample of the population? This may seem a somewhat pedantic question, but it is vitally important for the interpretation of the analysis. At some stage of the argument, it is almost certain that, explicitly or implicitly, someone, if not the analyst, will want to make the inference that, because the sample behaves in a particular way, the population behaves in the same way. Unless the individuals are taken from a defined population by some objective method of sampling, there may be very little value in undertaking multivariate analysis of the data.

Unfortunately, the apparent complexity of multivariate analysis seems to bemuse scientists into logical traps that they would readily avoid if they were analysing only one variable at a time. Most scientists are not well aware of the dangers of attempting to draw inferences from subjective samples, and the use of random or systematic sampling as a protection against subjective selection is well understood. It seems odd, therefore, that is should be necessary to assert that multivariate analysis is a statistical tool, and is subject to exactly the same logical constraints as any other statistical technique.

If, as is usually essential, the individuals to be included in the analysis are a representative sample from some defined population, there will be a 'structure', i.e. a sampling design or experimental design associated with the individuals which will represent independent samples from the population. In more complex cases, the population may have been stratified before sampling, so that the individuals come from several strata, and it will be necessary to identify the stratum from which each individual came. More rarely, the 'individuals' may represent 'plots' or a designed experiment, and, in such cases, it will be desirable to distinguish between 'blocks', or some other device to control experimental variability, and 'treatments' which may have a factorial structure.

The limit on the number of individuals that can be included in an analysis will depend on the forms of data storage available on the computer used for the analysis, but it is generally possible to analyse large numbers of individuals once the data have been summarised in some convenient way. However, the number of individuals will limit the use of particular methods of analysis in some instances; for example, cluster analysis is feasible only for relatively small sets of individuals, unless the analyst is prepared to spend a considerable amount of time (and money) on computing. Where data are available from very large numbers of individuals, some method of deriving a sample of individuals will often be desirable.

7.4.4 Construction of the basic data matrix

When the attributes and the individuals to be included in the study have been defined, the next stage in the analysis is to construct the basic data matrix. This matrix will have as many rows as there are individuals, and as many columns as there are attributes, with the values for each attribute given in a standard order for each individual.

Many analysts are tempted, in constructing the basic data matrix, to add to the measured variables additional sets of simple ratios of the measurements, or even to replace the variables by ratios constructed from them. This temptation should, however, be resisted. As Barraclough and Blackith (1962) have pointed out, 'the use of ratios implies certain prior knowledge about the nature of the systematic variation which, if set out explicitly, would almost certainly be denied by any experienced worker'. If a ratio hypothesis is preferred, there are simple ways of introducing the hypothesis at a later stage in the analysis.

Once the basic data matrix has been constructed, it will then be necessary to transfer the data to some machine-readable format. This format will depend on the input medium available on the computer to be used for the calculations. Increasingly, however, now that interactive computer systems are becoming more readily available, data are keyed direct into disk, magnetic tape or random access memory. This form of direct entry of data, under program control, is more efficient, as input errors can be detected rapidly and the use of programmed data editing and sorting enables error-free data to be obtained more quickly and more economically. The construction of the basic data matrix is therefore the stage at which the transfer should be made to a computer-readable medium of data storage. The point is emphasized because much valuable time can be wasted by attempting to do some of the early stages of the computations by hand or on programmable calculators. Not only can serious errors be introduced into the subsequent stages of the calculations, but it may be difficult or impossible to check the early calculations in any satisfactory way, so that a feeling of uncertainty is introduced into the later stages of the interpretation of the analysis.

7.4.5 Missing values

When the basic data have been assembled, it will frequently happen that there are missing elements in the two-way matrix of values, and the analyst will need to decide how to overcome the absence of these elements. There are no simple solutions to the problem. Most methods of multivariate analysis require complete sets of data for the calculations to be feasible, and some way will need to be found to eliminate the missing values or replace them in the basic data matrix.

The simplest solution is often to eliminate either rows or columns (or both) of the matrix so as to remove the missing values. This elimination needs to be done with care, however, as it is very easy to restructure the analysis so that it becomes irrelevant to the objectives of the research. The feasibility rows and columns will also depend on the number of missing values and the way in which they are distributed in the matrix.

If, for example, all of the missing values are concentrated on one column, representing a variable which was difficult to measure or which was not present on some individuals, consideration might be given to eliminating that variable from the subsequent calculations. Before doing so, however, it is important to ask 'Why was this variable included in the first place?' If it was included as one of a set of related variables thought to measure some particular dimension of variability, there may be little value in retaining it in the analysis: indeed, the fact that it has not been possible to obtain values of that variable for every individual has already indicated that it is not a very convenient measure of the variability that is to be investigated. On the other hand, the variable may be the only representative of an important dimension that must be included in the analysis if it is to meet the research objectives, and, in this case, one would be extremely reluctant to eliminate the variable if some other way of avoiding the missing value can be found. Where two or more variables have substantial numbers of missing values, any attempt to avoid the missing values by eliminating the variables may substantially restructure the whole analysis.

Similarly, if all, or most, of the missing values are concentrated in a few individuals, these individuals could possibly be eliminated from the analysis. Again, eliminating these individuals may not have any serious implications for the interpretation of the results, but might have the effect of changing the population from which the individuals are drawn. It is relatively easy to think of situations in which the difficulty of obtaining particular measurements is characteristic of certain parts of the total population, and elimination of the individual from that part of the population effectively changes the definition of the population from which the sample individuals are taken. The effects of eliminating individuals may be even more serious when there is an *a priori* structure imposed upon the individuals, as, for example, when the individuals represent 'plots' of a designed experiment. When variables and individuals have to be eliminated in order to provide a basic data matrix which is free of missing values, difficulties arising from the change in the research objectives may be particularly serious.

Difficult through the decision to eliminate variables, individuals, or both, from the analysis may be, reducing the basic matrix to a matrix without missing values may be the easiest solution that is available. Various attempts have been made to find a technique for replacing missing values in multivariate data, but none of the solutions so far devised are completely satisfactory. Where there is a definite structure that can be imposed upon the individual, any of the missing-value techniques appropriate to the experimental design may be used to fit small numbers of missing values for each variable separately. Alternatively, multiple regression analysis may be used to fit the missing variables, the prediction equations for the missing values being derived from those individuals for which complete data are available. Any missing values fitted in these ways must be examined very critically: it is better to reject an individual or variable than to include a 'wild-looking' missing value. In many situations, there may be little to be gained from anything but the replacement of the missing value by the mean of the variable for the other individuals.

Perhaps the most important question to ask, however, is the following: 'With these missing values, is any kind of multivariate analysis worth attempting?' The right answer may well be 'No'.

7.4.6 Scaling, standardization and transformation of the data

There are three ways in which multivariate data may be modified: (i) scaling; (ii) standardization; (iii) transformation. *Scaling* may be done in a variety of ways, the simplest being to add or subtract a constant from all values of a given attribute. Another method is to multiply or divide by a constant. *Standardization* implies that the value of each attribute for each individual is expressed as a derivative from the mean of that attribute, and then divided by the standard deviation. This procedure has the effect of reducing all attributes to unit standard deviation, and of reducing the magnitude of each attribute. Other methods include ranging (Gower, 1971) and rankits (Sokal and Rohlf, 1962). Standardization of attributes make them dimensionless, and so renders them additive. It also reduces the range, so that single attributes, or a small number of attributes, do not automatically dominate the analysis. The choice of the method of standardization is effectively a form of weighting.

The term *transformation* is used of methods which seek to change the shape of the frequency distribution of the data, usually in the hope of obtaining an approximately normal distribution. In univariate statistics, for example, transformations may be used to satisfy the theoretical requirement of the analysis of variance. Classical multivariate theory has been based largely on the multivariate normal distribution, and, although multivariate normality is not essential in some multivariate methods except for the sampling theory (e.g. in canonical correlation), not much is known about the robustness of the methods and the effects of large departures from normality. Some methods are sensitive to non-normality, e.g. cluster methods based on the assumption of a mixture of multivariate normal distributions. In numerical classification, dissimilarity measures may be sensitive to certain types of data. Thus, measures of Euclidean distance, including variance measures, are particularly sensitive to data in which there are occasional very large values (Clifford and Stephenson, 1975).

Transformation of the basic data may be introduced for the following purposes:

(i) to make the variables approximately normally distributed;

(ii) to make the sample variances and covariances of separate groups more homogeneous;
(iii) to make the relationships between the variables more nearly linear;
(iv) to simplify the interpretation of the analysis, e.g. to provide ratios of variables rather than linear combinations.

Although in practice it is difficult to separate these purposes, we may consider them in turn.

7.4.6.1 Transformation to make the variables approximately normally distributed

It may seem desirable to transform each of the variables included in the analysis so that they are all approximately normally distributed. So much of statistical theory is dependent on an appeal to the normal distribution that the requirement will sometimes seem to be almost a standard response to the analysis of data. Almost certainly, variables will occur in data matrices which cannot, by any shade of imagination, be regarded as even approximately normally distributed. The emphasis on the multivariate normal distribution in most texts dealing with multivariate analysis has the effect of misleading many users of multivariate techniques into thinking that the assumption of a normal distribution is essential for any practical application.

In fact, very little appeal to the multivariate normal distribution is necessary in the application of most forms of multivariate analysis. The essential mathematics of multivariate analysis does not depend of the assumption of the multivariate normal distribution, unless tests of significance are invoked. Such tests of significance are not usually required for the interpretation of the data.

Certainly, however, there may be little harm in introducing various transformations in order to make the distributions of the variables more nearly normal. The most frequent transformation for this purpose will usually be the logarithmic transformation, but the square root transformation may sometimes be suggested for variables which follow an approximate Poisson distribution, especially where all the values are less than 25, and the $\sqrt{(x+\frac{3}{8})}$ transformation may be used when all the values of one variable are less than 10. The arc sin transformation ($y = \sin^{-1}\sqrt{p}$) may be useful when a variable consists of percentages of a fixed number.

However, the value of transformations for the purpose of making the variables approximately normally distributed may be severely limited, quite apart from the fact that the assumption of a multivariate normal distribution is seldom of practical value in the interpretation of the results. The multivariate normal distribution places constraints on the correlation between the variables, and these will usually be affected in unknown ways by the transformations.

7.4.6.2 Transformation to make the sample variances and covariances more homogeneous

It is frequently desirable to reduce the heterogeneity of the sample variances and covariancs, especially in canonical variate analysis, where it is necessary to pool the 'within-group' variances and covariances. Again, the logarithmic transformation will be the one most frequently used for this purpose, although other transformations are also used regularly, including the square root, arc sin, and inverse transformations. It is worth checking that the proposed transformation will have the desired effect, as it is relatively easy to introduce some distortion into the results by the unwise use of transformations for one purpose.

7.4.6.3 Transformation to make the relationships between the variables more nearly linear

Most methods of multivariate analysis assume linear combinations of variables, and are hence more appropriate for use in situations in which the relationships between the variables are linear rather than curvilinear.

There are occasions on which the use of a transformation for one or more variables will greatly improve the linearity of the relationships between the whole set of variables. Again, logarithmic transformations are the most frequently invoked, and are often valuable when the variables show exponential or hyperbolic relationships. Note, however, that the use of transformations to make the variables more nearly linearly related also has effects on the sample variances and covariances, and these effects may be undesirable. Furthermore, while it may be easy to find a transformation which will linearize the relationship between one pair of variables, it may not be easy to find one that will linearize the relationships between all pairs (Kendall, 1975).

7.4.6.4 Transformation to simplify the interpretation of the analysis

For some applications, it may be desirable to introduce a transformation of the variables of the basic data matrix to simplify the interpretation of the results. One commonly occurring example of this kind of transformation has already been touched upon, i.e. when it is felt desirable to express the components of variation as ratios of the basic variables rather than as linear combinations of the variables. By transforming each of the variables of the basic matrix into their logarithm, the resulting linear combinations of the transformed variables which are provided by eigenvectors are complex ratios of the original variables. Such ratios may be easier to interpret than the linear combinations of the original variables, although, again, the logarithmic transformation has profound effects, which may or may not be beneficial on the variability of the variables and the linearity of the relationships between them.

The choice of whether or not to transform some or all of the variables of the basic data matrix is essentially subjective. In a sense, an analysis based on transformed variables represents an alternative hypothesis to the analysis based on untransformed variables, just as the choice of variables to be included in the analysis represents an hypothesis of the relevance of the variables.

Various types of transformation have been used in practical applications (Sneath and Sokal, 1973; Clifford and Stephenson, 1975). Andrews *et al.* (1971) have discussed the problem of transformations in which the transformed variables are functions of the original variables, collectively rather than separately, and have suggested some techniques which are occasionally useful. Clifford and Stephenson (1975) suggest that it remains uncertain whether the transformation required to produce normality of data is also the transformation which will produce optimal ecological 'sense', and that optimal ecological classificatory 'sense' is generally obtained by using a weaker transformation than that required to transform data to normality. In any multivariate analysis, careful thought needs to be given to the nature of the data and how they relate to the methods to be used. Despite any constraints, attributes will not, in general, contribute equally to the final classification. Any attribute which varies little over the population has little or no discriminating power.

7.4.7 Data exploration

Once the data have been assembled and stored on some medium accessible to the computer, they should be examined carefully as a preliminary step to any formal multivariate analysis. Methods for such data exploration have been presented by Tukey (1977), Mosteller and Tukey (1977), and McNeil (1977), among others. They include the calculation of medians, quartiles, means and standard errors, the identification of outliers and extreme values, plotting of histograms and boxplot diagrams to reveal skewness of distributions, and trial of transformations to correct various deficiencies. Plotting of two variables at a time on scatter diagrams helps to reveal non-linearity of the relationship between variables, and, again, transformations can be tried to correct for the non-linearity and heteroscedacity of the relationships.

All of these methods are now readily available on modern interactive computer systems, and are invaluable in revealing previously undetected errors in the basic data, as well as in increasing the general understanding and 'feeling' for the data. A present tendency in the use of computers to reduce the degree of contact between scientists and their data is regrettable and totally unnecessary. The modern, time-sharing, interactive computer or microprocessor offers, as never before, opportunities to explore data before embarking on any major analysis. In this way, it is possible to check that the assumptions in the use of particular methods of analysis are justified.

7.5 CHOICE OF METHOD OF ANALYSIS

There are now so many different methods of multivariate analysis described in the literature that it would be impossible to compile a complete catalogue of methods. Fortunately, all of the methods fall into a few major categories, so that it is sufficient to describe these categories and some of the more important techniques within each category. Anyone embarking on the use of multivariate techniques of model dynamic change in ecosystems would be well advised to begin with one of the more well-established techniques before moving to the many less well-tried methods.

The multivariate models presented in this chapter will be summarized in four main categories, namely ordination, discrimination, classification and the fitting of relationships.

7.5.1 Ordination

The basic $n \times p$ data matrix defines the position of the n individuals in a p-dimensional attribute space. Ordination procedures in multivariate analysis have as their aim the arrangement of the individuals in as small a number of dimensions as possible, while retaining as much as possible of the information contained by the data. The reduction in dimensionality from the total number of attributes to (hopefully) some smaller number makes the data easier to handle mathematically by: (i) replacing attributes which are linearly related, or nearly so, by a single composite attribute; (ii) making graphical representation easier by referring the data points to axes which are orthogonal, i.e. at right angles to each other; and (iii) increasing the possibility of reification, i.e. the interpretation of the mathematical expression of the data in terms of the original problem by providing a useful insight into their structure.

Ordination procedures are necessarily related to the amounts of information which are available about either the attributes or the individuals. Any cluster of attributes or individuals will need to be related to this information. But the analysis begins by assuming that the basic data matrix is unpartitioned, that there are no *a priori* divisions of this matrix which must be incorporated into the assumptions underlying the multivariate model, and that the external information will be determined after these parameters have been estimated. Figure 22 emphasizes this lack of *a priori* divisions of the matrix, by contrast with the figures which follow.

Individuals	*Attributes* 1	2	3	...	p
1	x_{11}	x_{12}	x_{13}	...	x_{1p}
2	x_{21}	x_{22}	x_{23}	...	x_{2p}
3	x_{31}	x_{32}	x_{33}	...	x_{3p}
⋮	⋮	⋮	⋮		
n	x_{n1}	x_{n2}	x_{n3}	...	x_{np}

Figure 22. Undivided matrix for ordination models

7.5.2 Discrimination

The purpose of discriminant analysis is to examine how far it is possible to distinguish between members of various groups on the basis of observations made about them (Marriott, 1974). Thus, the basic data matrix is divided, *a priori*, into two or more groups. The multivariate model for discrimination provides for:

(i) tests of significance for differences in the values of the attributes between the groups;
(ii) allocation rules for identifying further individuals as belonging to one of the groups by some kind of discriminant function based on the measured attributes;
(iii) estimates of the probability of correct allocation by the rules that are derived from the model.

As shown in Figure 23, the basic data matrix now has divisions of the individuals into two or more groups, based on information which is external to the analysis itself. It is not necessary that there should be the same number of individuals in each group.

Individuals		*Attributes* 1	2	3	...	p
	1	x_{11}	x_{12}	x_{13}	...	x_{1p}
Group 1	2	x_{21}	x_{22}	x_{23}	...	x_{2p}
	3	x_{31}	x_{32}	x_{33}	...	x_{3p}
Group 2	⋮	⋮	⋮	⋮		
Group r	⋮	⋮	⋮	⋮		
	n	x_{n1}	x_{n2}	x_{n3}	...	x_{np}

Figure 23. Division of basic matrix into r *a priori* groups for discriminant analysis

7.5.3 Classification

Multivariate models for classification or cluster analysis fall into four sub-categories, which overlap to a considerable extent.

(i) What is the best way of dividing the individuals into a given number of groups, a procedure called 'dissection' by Kendall and Stuart (1979)? The resulting dissection need only be a convenient way of dividing the given set of individuals into groups, as opposed to some 'natural' division of the data.

(ii) What is the best way of dividing the given set of individuals into more or less homogeneous and distinct categories? In ecology, we often expect a set of individuals to reflect biological characteristics which enable organisms or communities to be classifiable into groups which are in some way 'natural'.

(iii) Are there discontinuities in the multivariate attribute space in which the individuals are described which will help us to classify those individuals? Here, we are content to be guided by the analysis in the identification of distinct groups in the space defined by the attributes.

(iv) What is the best way of constructing a strictly hierarchical classification and plotting a dendrogram for these individuals? Expressing the relationship between the individuals in the form of a tree implies a special kind of structure, and is frequently confused with a search for natural groupings.

In practice, these four sub-categories are mixed up in any single application of multivariate models. A very large number of possible methods have been reviewed by Cormack (1971). The choice of an appropriate method therefore becomes quite difficult, and only a few basic approaches will be described in this Handbook.

The division of the basic data matrix is the same as in Figure 23, but the individuals must now be re-ordered. The divisions of the individuals into groups are not made on the basis of external information, but solely on the basis of information contained in the data matrix. The ordering of the individuals within the matrix will, however, usually need to be changed to reflect the groupings or dissections which are identified by the analysis.

7.5.4 Relationships between variables

The fitting of defined relationships between a single variate, y, and a set of variables, $x_1, x_2, \ldots, x_p$, which may or may not be random variables, is a well-known problem usually solved by the technique of multiple regression. Strictly speaking, multiple regression is not a multivariate model, but is a technique so widely used, and often misused, that a brief account of it will be given later in this chapter. The multivariate extension of the concept to the relationship of a set of variates with another set of variables is of particular importance in the modelling of dynamic change in ecosystems, but it is an extension which has been somewhat neglected in ecological research. The technique defines the relationship between any two sets of attributes by pairs of newly defined variables. It is then possible to test whether there is evidence for any relationship, whether the relationship is accounted for entirely by the first pair, or the first few pairs, of variables, and whether some of the original variables can be left out of the consideration, without significantly affecting the conclusions (Marriott, 1974).

As shown in Figure 24, the divisions of the basic data matrix now occur between sets of variables. In a complex case, there may be several groups of variables, but the appropriate model for this case is an extension of the simple model case for only two groups of variables. Ideally, both sets of attributes should consist of sets of continuous variables, but modifications of the technique exist for other types of variables. Again, it is not necessary that the two groups of attributes should be of the same size.

The choice between these four main categories necessarily depends on the purpose of the investigation. In part, the choice also depends on how much is known about either the attributes or the individuals or both. In other words, the multivariate model is embedded in a whole series of hypotheses, assumptions and objectives. Unfortunately, this fact is seldom made clear in any of the texts on multivariate analysis, with the result that the basic data matrix is usually regarded as being self-sufficient in representing all that is known about the problem the multivariate model is being used to solve. This section of the Handbook on the choice between methods is, therefore, intended to emphasize the importance of the information which is external to the data matrix itself. Descriptions of some of the more important techniques within each of the main categories are given in the following sections.

Individuals \ *Variables*	1	2	3	...	r	s	...	p
1	x_{11}	x_{12}	x_{13}	...	x_{1r}	x_{1s}	...	x_{1p}
2	x_{21}	x_{22}	x_{23}	...	x_{2r}	x_{2s}	...	x_{2p}
3	x_{31}	x_{32}	x_{33}	...	x_{3r}	x_{3s}	...	x_{3p}
$\vdots$	$\vdots$	$\vdots$	$\vdots$		$\vdots$	$\vdots$		$\vdots$
n	x_{n1}	x_{n2}	x_{n3}	...	x_{nr}	x_{ns}	...	x_{np}

Figure 24. Division of basic matrix into two groups of variables for fitting of relationships between the two groups

7.6 METHODS OF MULTIVARIATE ANALYSIS: INTRODUCTION

The description of the underlying mathematics of multivariate techniques in this Handbook is restricted to the information which the more mathematically inclined reader may find useful in relating the methods applied to the practical analysis of data to algebraic theory. There are plenty of alternative presentations of the mathematical theory of multivariate analysis, and anyone who wants to drink deeply from this particular source of inspiration would do better to read these other texts, and notably Anderson (1958), Hope (1968), Morrison (1967), Seal (1964), and Kendall and Stuart (1979). For the more limited purposes attempted here, the excellent paper on the algebraic basis of the classical multivariate methods by Krzanowski (1971) has been used as both a source of more or less direct reference and a consistent notation. An understanding of matrix algebra is essential for any compact presentation of the mathematics of multivariate analysis, and readers requiring comfort and instruction on this subject should consult Searle (1966) or Bellman (1960).

The essential nature of multivariate data has already been illustrated in Figure 21. We suppose that p attributes are observed on each individual in a sample of n such individuals. Generally, the observed attributes will be variates, i.e. they will be values of a specified set with a specified relative frequency or probability. In matrix terms, the values given to the variates for the ith individual are the scalars x_{ij} where j can take all the values from 1 to p. The whole series of values for the ith individual is given by the vector:

$$\mathbf{x}_i' = (x_{i1}, x_{i2}, x_{i3}, \ldots, x_{ip})$$

The complete set of vectors represents the data matrix $\mathbf{X}$ with n rows and p columns:

$$\mathbf{X} = (\mathbf{x}_1, \mathbf{x}_2, \mathbf{x}_3, \ldots, \mathbf{x}_n)$$
$$\mathbf{X}' = (\mathbf{x}'_1, \mathbf{x}'_2, \mathbf{x}'_3, \ldots, \mathbf{x}'_p)$$

and is a compact notation for the whole sample. The ith row of the matrix gives the values of the p variables for the ith sample: the jth column of the matrix gives the values of the jth variable for each of the individuals in the sample. Note that, at this point, no assumptions are made about the extent to which the sample can be regarded as representative of some defined population.

This compact notation for the data matrix has some convenient computational properties. Without loss of generality, we may assume that the variates are measured about their means for the sample, so that all the column means are zero. Then, the sample variance–covariance matrix may be calculated as:

$$[1/(n-1)]\mathbf{X}'\mathbf{X}$$

where $\mathbf{X}'$ is the transpose of the original data matrix. The sample sums of squares and products, $\mathbf{X}'\mathbf{X}$, and hence the variance–covariance matrix, has the mathematical property of being real, symmetric and positive semi-definite.

Geometrically, the sample data of Figure 21 can be represented as n points in p dimensions, with the values of the jth variate ($j = 1, 2, 3, \ldots, p$) for each unit referred to as the jth of p rectangular co-ordinate axes. When the number of variates is large, the resulting geometric representation is in many dimensions and cannot easily be visualized. For this reason, many of the techniques of multivariate analysis seek to simplify this representation by reducing its dimensionality, but also to keep the essential features of the data as measured by their total variance.

7.7 ORDINATION

In this section, four methods of ordination will be described, depending on the main focus of interest in the undivided basic data matrix. Where the main focus of interest, initially at least, is on the correlations between the variables, *principal component analysis* may be the most appropriate method of deriving an ordination and reification of the data. *Factor analysis* is regarded by some analysts as an alternative. Where, however, the basic data matrix is used to define the distances between individuals in the attribute space, *principal co-ordinate analysis* finds the n points relative to principal axes which will give rise to these distances. Finally, an iterative approach may be used to derive an ordination simultaneously from the attributes and the individuals by *reciprocal averaging*, a technique which is related to *correspondence analysis*.

7.7.1 Principal component analysis

Where no *a priori* structure is imposed on the data matrix, so that it represents a scatter of n points in p dimensions, we seek a rotation of the axes of the multivariate space such that the total variance of the projections of the points on the first axis is a maximum. We then seek a second axis orthogonal to the first, which accounts for as much as possible of the remaining variance, and so on.

If $\mathbf{x}' = (x_1, x_2, \ldots, x_p)$ represents a point in the p-dimensional space, the linear combination $\mathbf{l}'\mathbf{x}$ of its co-ordinates represents the length of an orthogonal projection on to a line with direction cosines $\mathbf{l}$, where $\mathbf{l} = (l_1, l_2, \ldots, l_p)$ and $\mathbf{l}'\mathbf{l} = 1$.

The sample variance of all n elements is given by:

$$V = \mathbf{l}'\mathbf{X}'\mathbf{X}\mathbf{l}$$

and to maximize V subject to the constraint of orthogonality, the following criterion is maximized:

$$V' = \mathbf{l}\mathbf{X}'\mathbf{X}\mathbf{l} - \lambda(\mathbf{l}'\mathbf{l} - 1)$$
$$= \mathbf{l}'\mathbf{W}\mathbf{l} - \lambda(\mathbf{l}'\mathbf{l} - 1)$$

where $\mathbf{W} = \mathbf{X}'\mathbf{X}$.

It can be shown that the p equations in p unknowns $1, 2, \ldots, p$ have consistent solutions if and only if $|\mathbf{W} - \lambda\mathbf{I}| = 0$. This in turn leads to an equation of degree p in with p solutions $\lambda_1, \lambda_2, \ldots, \lambda_p$. These solutions are variously designated as the *latent roots, eigenvalues* or *characteristic roots* of $\mathbf{W}$.

Substitution of each solution $\lambda_1, \lambda_2, \ldots, \lambda_p$ in

$$(\mathbf{W} - \lambda\mathbf{I})\mathbf{l} = 0$$

gives corresponding solutions of $\mathbf{l}$ which are uniquely defined if the λ's are all distinct, and these are designated as the *latent vectors, eigenvectors*, or *characteristic vectors* of $\mathbf{W}$.

The extraction of the eigenvalues and eigenvectors of the variance–covariance matrix of our original data matrix representing n points in p dimensions neatly defines the linear combination of the original variables which account for the maximum variance while remaining mutually orthogonal. The elements of the eigenvectors provide the appropriate linear weighting for the components, and the eigenvalue, expressed as a proportion of the number of dimensions (p), gives the proportion of the total variance accounted for by the component.

Note that, in the argument used so far, no assumptions about the normality of the multivariate distribution from which the samples have been drawn have been involved, and such assumptions would only be required if tests of the significance of components were required.

7.7.1.1 Covariances or correlations?

The above theory has been developed for the covariance for the covariance matrix and is most suitable if all the original variates have been measured in the same units. Even if the variates are all on comparable scales, it is necessary to be careful in working with unstandardized data. For example, if the range of values obtained for one variate is greatly different from that of another, it clearly makes a good deal of difference whether the data are standardized or not. On the other hand, if the original variates are not in the same units, their linear compound would have little meaning, and the rationale of maximizing $\mathbf{a}'\mathbf{S}\mathbf{a}$ relative to $\mathbf{a}'\mathbf{a}$ is questionable; in fact, the analysis will depend on the different units of measurement (Anderson, 1958).

If the variates are not all expressed in the same units, it is customary to standardize the data so that:

$$z_{ij} = x_{ij}/s_j$$

where x_{ij} is an element of matrix **X** and is a deviation from the mean, and s_j is the standard deviation of the jth variate. In this case, the transformation is:

$$\mathbf{U} = \mathbf{ZB}$$

and the covariance matrix **U**, say **L**, is:

$$\mathbf{L} = \mathbf{B}'\mathbf{RB}$$

where **R** is the covariance matrix of **Z**, and the correlation matrix of **X**.

The theory for principal component analysis based on the correlation matrix is essentially the same as that for the covariance matrix (see Cooley and Lohnes, 1971). The eigenvalues l_j and eigenvectors $\mathbf{b}_j$ of the correlation matrix are computed. The sum of the eigenvalues will be the trace of **R**, and will be equal to p. In analysing the correlation matrix, the data are regarded as a set of correlated variates, and the aim of the linear transformation is to obtain an equivalent set of uncorrelated components with maximized variances. Hence, the pattern of the eigenvector elements depends upon the correlation structure of the observations. In an analysis of the covariance matrix, the pattern of the eigenvector elements will depend on the variance–covariance structure.

7.7.1.2 Eigenvalues and eigenvectors

The eigenvalues and eigenvectors of a covariance or correlation matrix determine the lengths and directions of the component axes. The orthogonal eigenvectors used in computing the component values are the direction cosines of the component axes. The columns of an orthonormal eigenvector matrix are the eigenvectors, and the rows represent variables. As the sum of the squared elements of any column is unity, the square of any element gives the proportion of the variance of the component which is accounted for by the corresponding attribute. The contribution of a component to an attribute is not so obvious, and, by concentrating only on the orthonormal matrix, the possibility may be overlooked that a particular attribute is largely accounted for by one of the later components. In order to get more information, other normalizations of the eigenvectors are necessary.

The eigenvectors of a covariance matrix may be normalized so that the sum of the squared elements of a column eigenvector is equal to the corresponding eigenvalue. Then, providing that all the eigenvectors are included in the matrix, the sum of the squared elements in any row will equal the variance of the corresponding attribute. The elements of this matrix are $a_{ij}\sqrt{d_j}$. For convenience, we can call this a type A eigenvector matrix. The best fitting q-dimensional subspace to a scatter of points in a higher (p) dimensional space passes through the centre of gravity of the points, and the sum of squares of the perpendiculars from the points to the subspace is a minimum. Such a space is defined by the centre of gravity (i.e. the mean vector) and the orthonormal eigenvectors. Any point in the subspace can be referred to the p original axes by the relation:

$$\mathbf{x}_i = \mathbf{x}' + \mathbf{k}'_i\mathbf{C}'$$

where $\mathbf{x}_i$ is a p-dimensional row vector of the raw data matrix $\mathbf{X}$, $\mathbf{x}'$ is the mean vector, $\mathbf{k}'_i$ is a row vector of a matrix $(N \times q)$ of coefficients, and $\mathbf{C}'$ is the $(q \times p)$ transpose of:

$$\mathbf{C} = \mathbf{AG}$$

where $\mathbf{G}$ is a diagonal matrix of square roots of the eigenvalues, i.e. matrix $\mathbf{C}$ is a type A eigenvector matrix. The elements of the eigenvectors:

$$c_j = a_j d_j$$

are 'typical points', summarizing the deviation of the original points from the centre of gravity (Rao, 1964). In some types of study, the typical points prove to be useful (e.g. Simonds, 1963). In other types of study, they may not admit any useful interpretation. Taylor (1977) has plotted these values as 'variance profiles'. Each observed profile, easier to interpret if the data are standardized, is a composite of several such profiles.

If each element of a row of a type A eigenvector matrix is divided by the square root of the variance of the corresponding variable, the rows will have unit sum of squares. Squaring the appropriate element gives the proportion of the total values of an attribute which is accounted for by a particular component. The elements will be $a_{ij}\sqrt{d_j}/s_j$, i.e. the product–moment correlation of the ith attribute and the jth component. Each element is a cosine between an attribute and a component, i.e. a direction cosine of the attribute. It is also a projection of the attribute on to a component (Hope, 1968).

An eigenvector matrix obtained from a correlation matrix bears no simple relation to that derived from the corresponding covariance matrix. Here, too, the orthonormal eigenvectors $\mathbf{b}_j$ are used to compute the component values, if they are applied to the standardization data. The eigenvector can also be normalized so that the sum of the squared elements of a vector equals the associated eigenvalue, the new vector elements being b_{ij}. The sum of squares of each row is unity, and squaring the appropriate element gives the proportion of the total variance of an attribute which is accounted for by a particular component. Each of these elements is the correlation of the ith attribute with the jth component, and they are direction cosines of the attributes. For convenience, we can call this a type D eigenvector matrix.

Finally, for both a covariance and a correlation matrix, the elements of an orthonormal eigenvector matrix can be divided by the square root of the corresponding eigenvalue (Hope, 1968; Massey, 1965). If such a matrix is applied to the data, the resulting components will have unit variance. We can call this a type C matrix (covariance) or a type E matrix (correlation).

7.7.1.3 Calculation of eigenvalues and eigenvectors

The main computational problem of principal component analysis is to solve the eigenstructure of a symmetric matrix. Since the use of digital computers became widespread, significant advances have been made in the development of fast and accurate algorithms. Modern methods first reduce the matrix to tridiagonal form by Householder's method. The eigenvalues of the tridiagonal matrix are the same as those of the original matrix. The eigenvectors are not the same, but are readily found by an appropriate transformation. In the algorithm given by Ortega (1967), the eigenvalues are calculated by the method of Sturm sequences. As many eigenvalues as required may be calculated, but it is usual to compute all the eigenvalues, if not all the eigenvectors, because the sum of the eigenvalues gives the rank of the data matrix. Finally, the eigenvectors of the tridiagonal matrix are calculated and back-transformed to eigenvectors of the original matrix by a procedure due to Wilkinson (1965).

In the above method, if the matrix was pathologically close to being singular, or if multiple eigenvalues exist, the eigenvalues will be accurate but the eigenvectors will not be orthogonal. If accurately orthogonal eigenvectors of close or multiple roots are required, it is better to use the TRED2 routine of Martin *et al.* (1971) for the tridiagonal matrix and the QL algorithm with standard shift (TGL2 routine of Bowdler *et al.*, 1971) to calculate the eigenvalues and eigenvectors. This combination produces eigenvectors which are always accurately orthogonal, even for multiple roots. The order in which the eigenvalues are found is, to some extent, arbitrary. The accuracy of individual eigenvectors is, of course, dependent on their inherent sensitivity to changes in the original data.

When the complete set of eigenvalues and eigenvectors is required, Ortega's program is more efficient than TRED2 and TGL2 for matrices of orders exceeding about 48, but TRED2 and TGL2 are more efficient for orders less than about 35 (Sparks and Todd, 1973). TRED2 and TGL2 also have the advantage of requiring less storage space.

7.7.1.4 Dimensionality of the data matrix

Principal component analysis normally provides $r \leqslant p$ non-zero eigenvalues from p attributes. If $n < p$, a maximum of $n - 1$ components are obtained. For many purposes, it is convenient to disregard components which have small variances, treating them as constants. Hence, the main components of variation may be studied in a subspace of dimension $q < p$. Although some information is lost in this transformation, the q components may account for enough of the original variance to make the reduced dimensionality useful. Indeed, much of the variability in the lower components may be regarded as random noise. However, because components are rejected the possibility should be considered that one of the lower components may be important to a particular attribute or to a particular subgroup of individuals. In choosing to ignore a component, we are not doubting its reality. Any dimension which exists in a sample must exist in the population, and, however small the latent root, it should not have arisen from a population in which the corresponding value is zero (Kendall, 1975). We are merely suggesting that that part of the variance is not relevant to the particular study in hand (Jolliffe, 1982).

For some, data, the first few eigenvalues are large, but successive eigenvalues decrease sharply. It is then clear how many components are likely to be of practical importance. In many cases, however, the result is less clear. The first component may account for half, or less, of the total variance, and successive eigenvalues decrease gradully. This result may be due to one of two causes. First, the variances of the original measurements are approximately equal (or have been made to be). Second, the correlation between the original measurements may be strongly curvilinear (Williams, 1976).

Morrison (1976) suggests that, from his experience, there is usually little point in extracting further vectors if the first four or five components do not account for about 75% of the variance. Even if the later eigenvalues were sufficiently distinct to allow easy computation of the components, the interpretation of those components may be difficult or impossible. He suggests that it is frequently better to summarize the data in terms of the first components with large markedly distinct variances, and to include as highly specific and unique variates those responses represented by high loadings in the later components, although the latter are likely to be associated with considerable noise from the other, unrelated variates. Marriott (1974) suggests that, if the attributes have been standardized, it may be reasonable to discard all components with a smaller variance than that of a single attribute, but, on the whole, it is better to retain unimportant components, than to discard information of value.

Jeffers (1965) proposed a 'rule of thumb' test based on the fact that, if the correlation matrix is a unit matrix, i.e. no attribute correlates with any other attribute, the eigenvalues will all be unity. If there are correlations, some of the eigenvalues will be greater than unity and some will be less. Hence, a component with an eigenvalue less than unity represents a component which accounts for a smaller proportion of the variance than would be represented by each of the basic attributes separately. In practice, therefore, he has found it useful to consider only those components having eigenvalues equal to or greater than unity, although he might also look at the next one or two components with eigenvalues less than unity, provided that they are greater than about 0.75. Kaiser (1960) has also expressed a variety of arguments for a 'rule of thumb' test based on eigenvalues greater than unity, and this rule seems to work well with small or moderate-sized samples (Cooley and Lohnes, 1971).

7.7.1.5 Sphericity

A different question is that of sphericity. The well-known chi-square test of Bartlett, with its various modifications (Kendall, 1975), tests whether some or all of the eigenvalues are equivalent. Applied to all the eigenvalues, the test has different interpretations according to whether it is applied to the correlation matrix or the covariance matrix. Applied to the correlation matrix, the test asks whether the eigenvalues show any tendency to deviate from a uniform value of unity. If they do not, the correlations among the tests cannot be assumed to differ from zero, and there is no point in transforming the data to orthogonal components. If the test is applied to the covariance matrix, it is a test of the independence of the variates and of the equality of their variances (Hope, 1968).

Such statistical tests require a large sample drawn from a p-dimensional multivariate normal population with a covariance matrix having p distinct, non-zero eigenvalues. The asymptotic distribution theory of the eigenvalues and eigenvectors of correlation matrices is considerably more complicated than that of covariance matrices (Anderson, 1963). Furthermore, it does not follow that all the components which reach statistical significance in a large sample necessarily remove a large proportion of the variance, and so some of them may be comparatively unimportant in practice (Bartlett, 1950). Krishnaiah and Lee (1977) have discussed the problem of the testing of roots of covariance matrices. James (1977) described tests of the hypothesis that a prescribed subspace is spanned by principal components. The value of such tests in the analysis of real data is doubtful, and many workers prefer to use the 'rule of thumb' approach given above. Non-parametric tests might prove useful, but do not seem to have received much attention.

7.7.1.6 Calculation of the transformed values

As a further aid to the interpretation of the analysis, and as an input for further analysis and interpretation, it is usually desirable to compute the value of each of the components regarded as meaningful for each of the individuals in the basic data matrix. Where, as will most usually be the case, the components will be computed from the standardized variables of the basic matrix, i.e. from the difference of the individual from the average for the matrix, divided by the standard deviation of the values, the product of the weighted eigenvectors and the standardized variables for the individuals then gives the transformed values of the original data.

If the analysis has succeeded in giving any effective reduction of the essential dimensions of the problem, there will be fewer transformed variables than there were variables in the basic matrix, and these transformed variables have the desirable property of being orthogonal or statistically independent.

7.7.1.7 Analysis of transformed values

Many types of analysis can be carried out on the transformed values represented by the calculated components for each individual, and are limited only by the ingenuity of the analyst, and the purpose of the analysis. Much will depend on the structure that has been imposed on the variables and individuals of the basic data matrix. Where no such structure exists, some kind of cluster analysis (see below) of the individuals will usually be the only kind of further analysis which will seem appropriate, but, even here, the choice of methods is considerable. However, it will be evident that a preliminary component analysis of the data will frequently greatly reduce the dimensions over which the clustering needs to be performed. It will also remove much of the 'noise' which makes some forms of cluster analysis difficult to interpret.

Structure imposed on the variables will frequently lead to comparisons of the correlations between sets of components or to the calculation of the regression of a component of one set on all or some of the components of another set. Clearly, there is no point in looking at the correlations between components of the same set, as these are, by definition, orthogonal. Regression calculations can be greatly facilitated by the technique of orthogonalized regression described by Kendell (1975). Structure imposed on individuals will usually lead to analysis of variance of the individual components, the analysis being related to the experiment or survey design implied by the structure. The orthogonality of the components greatly increases the ease of interpretation of this method of analysis.

7.7.1.8 Plotting of transformed values

Plotting of the transformed values will frequently be helpful in making clear the relationships and distinctions between the variables, components and individuals. The most usual form of plotting will be as a projection of the n-dimensional space on to the plane formed by the axes of two of the components. The orthogonality of the components ensures that the axes of any pair of components are at right angles to one another, and greatly simplifies the plotting. However, it must always be remembered that projection of more than two dimensions on to a two-dimensional plane almost inevitably results in some distortion of the relationships between individuals. The techniques of minimum spanning tree and nearest neighbour analysis may be used to alert the analyst to the degree of distortion present.

Where plotting of three dimensions is required, several techniques are available, including the use of various kinds of symbols to indicate deviations from the two-dimensional surface in some third dimension, and the use of stereo pairs of plots giving the illusion of three dimensions when viewed through a simple stereoscope. For larger numbers of dimensions, Andrews' transformation (Andrews *et al.*, 1971) will usually be preferable. Alternatively, biplot graphical displays (Gabriel, 1971) may be used to obtain a visual appraisal of the structure of large data matrices, and to show the inter-unit distances and the clustering of the individuals in multivariate space. This is one of the areas of application of computer graphics which promises to greatly improve interpretation and understanding of complex data.

7.7.1.9 Reification

Reification is the interpretation of the results of a mathematical analysis in terms of the original problem and observations. Such of the components can sometimes be interpreted in terms of features of the original observations. Components may correspond to features which have already been appreciated in an intuitive way. It is important to realize the limitations of this type of reification. The order in which the components appear, the proportion of the variance associated with each, and, hence, the conclusions drawn, depend on the objects chosen and the observations included in the analysis, as well as on sampling variations. The stability of the components of a sample can be examined by making similar sets of measurements on several samples, and doing a separate analysis on each set. Kendall (1975) emphasizes that the direction cosines are sensitive to fluctuations such as one would get from one sample to another. It is therefore unwise to place too much emphasis on the numerical value of any particular coefficient in an eigenvector.

In some types of analysis, the first principal component has the character of a size vector, while others are shape vectors. Jolicoeur and Mosimann (1960), in studies on the painted turtle, emphasized that, for the first component to be a size vector, all coefficients must be of the same sign. Rao (1964) gave a mathematical justification for this conclusion. Examples of other studies of this type are given by Blackith and Reyment (1971). However, the first component should not be regarded as a size component unless the structure of the observations clearly suggests such an interpretation. For example, in studying heath vegetation (species presence–absence), Ivimey-Cook and Proctor (1967) found that the first component reflected species abundance.

In some applications, the calculated components do not suggest a simple reification, i.e. the pattern of eigenvector elements on a particular component is not susceptible to a simple biological or ecological interpretation. Holland (1969) has pointed out that, while the principal components themselves may be of little biological significance, the space defined by their vectors is significant, and it is only a matter of geometrical manipulation to determine the extent to which the vectors of other components corresponding to biological hypotheses, or derived from other bodies of data, lie within such a space. Holland further suggests that, if certain hypothetical or consistent components can be found to fit the observations, it may be preferable to define the space occupied by the observations in terms of these new components. If this is done, two problems arise. First, the variation associated with each new component will be unknown. Second, the number of such components may not be sufficient to define the space. He suggests ways of overcoming these problems.

7.7.1.10 Rotation

Various methods have been proposed for rotating the component axes to give a new set of axes spanning the same space. The proportion of the total variance accounted for by the first q components will be unchanged, but the variances of the individual new components will be different from the originals, and they will not be subject to maximum-variance constraints, i.e. they will not be the principal axes of the hyper-ellipsoid. Furthermore, the rotated axes may not be orthogonal, so that the variances of the new components do not constitute a partition of the total variance attributable to the component space. Hence, great care is needed in interpreting such components. If axes are to be rotated, the aims of the rotation must be clearly stated and formulated in mathematical terms, so that the angle of rotation can be determined. In many studies, ecologists apply one or more of the standard rotation techniques used in factor analysis (see Section 7.7.2 below).

For example, Ivimey-Cook and Proctor (1967), studying heath vegetation, found that, in principal component analysis of the correlation matrix, the first three components reflected species abundance, soil moisture and base status respectively. The fourth and fifth components were not readily interpretable. The first five components accounted for nearly 88% of the total variance. Examination of the ordination graphs suggested that a rotation of the axes might bring them into line with configurations of points, and the components were then rotated according to Kaiser's varimax criterion to give five new axes which were readily interpretable as corresponding to five recognizable vegetation types.

Another method of rotation involves the fact that, in a vegetation association which is virtually a pure stand, the entropy of mixing of the constituent elements is low relative to that of an association where many elements are approximately equally represented (Pelto, 1954). Zones of inter-gradation between two associations or communities have high entropy (Howard and Murray, 1969) and, if components are rotated to positions such that the entropy of the system is minimized, the resulting new axes might be expected to link the centres of associations (McCammon, 1966, 1968).

7.7.1.11 Interpretation of results

Having reviewed the main stages of principal component analysis, it will be evident that the interpretation of the results of the analysis depends on careful oversight of all these stages. It is perhaps for this reason that the interpretation of multivariate data appears to present so many problems to the inexperienced analyst. Used to less rigorous applications of statistical methods, the analyst may be in the habit of delegating some stages of the analysis to others, with a subsequent loss of control of the analysis and of the understanding of the results. Worst still, the analyst may be tempted to surrender the decisions that he should make for himself to the 'control' of a computer program or package — he may not even know what decisions have been programmed into the package, with disastrous results. It is the author's view that the correct interpretation of multivariate analysis can only be achieved if the data, and all the stages of the analysis, are properly embedded in the logic and background of the investigation, of which the data form part.

7.7.1.12 Advantages and disadvantages of principal component analysis

The main advantage of principal component analysis lies in the robustness of the least squares approach to approximating the covariance or correlation matrix. It is, therefore, not important for the data to be multivariately normal, unless significance testing is required. Other advantages lie in the relative simplicity of the technique. For example, it is easy to see the contributions made by the attributes to each component.

The main disadvantage lies in the assumption that any relationships among the original attributes are essentially linear, or at least that any non-linear contribution is small. Problems occur if the data are markedly non-linear, for example along an environmental gradient. The orthogonality of the components implies functional independence only if the objects are normally distributed. Norris (1971) used a simple example with artificial data to show that, if one attribute has a linear response to an environmental gradient while a second has a sinusoidal response, the resulting plot of the components shows the environmental gradient to be curved. Situations in which some of the attributes do not increase or decrease linearly along an environmental gradient proposed as a reification should be examined for functional relationships between components (e.g. Norris and Barkham, 1970).

Problems can occur with plant species, which may be thought of as being distributed along environmental gradients with frequency curves which are S-shaped if the peak value is at one end of the gradient, or bell-shaped if the gradient encompasses the whole range of a species along a gradient, as described by Whittaker (1967). If only a small part of the gradient is examined, the relationships may be approximately linear for a number of species, but, as more of the gradient is included, the number of zero occurrences increases and the bell-shaped curves overlap. Noy-Meir and Austin (1970) subjected a set of simulated data of this type to analysis. The resulting ordination diagram showed that the linear gradient became a complex three-dimensional curve. However, they concluded that, provided no assumption of a one-to-one relationship between axes and gradients is made, principal component ordination could still give a useful representation of the data.

Austin and Noy-Meir (1971) examined the problem of non-linearity using artificial data for two-gradient models. They distinguished two types of distortion which can arise: (i) involution, where extreme stands occur closer to the centre of the environmental plane than less extreme stands, and (ii) spurious axes which appear in the ordination although they do not represent independent environmental gradients.

A related problem concerns the nature of the matrix from which the eigenvalues and eigenvectors are calculated. Principal component analysis is a method for partitioning variances, and can be applied to sums of squares and products matrices, although it is more commonly applied to covariance and correlation matrices. Care is necessary when applying the method to correlation matrices which are not calculated by ordinary product-moment methods (Kendall, 1975). Binary data present special problems; the frequency of occurrence of an attribute constrains the range of correlation possible. The type of species distribution described above means that a species may contribute to the calculation of ecological distance in only a limited part of the data set, because it is represented elsewhere by zeros.

Where the linearity constraint of principal component analysis has received much attention, there has been less concern with the problems raised by the use of mixed data, and in the additive derivation of component values. Some workers think that multiplicative models might be more appropriate. Various problems and approaches to the application of principal component analysis were discussed by Dale (1975).

A problem which often occurs in multivariate studies is how to reduce the number of original attributes which need to be measured. Jolliffe (1972, 1973) discussed eight methods, four of which use principal components. He applied five out of eight techniques to real data, including a multiple linear regression analysis. Mansfield *et al.* (1977) presented a method for reducing the number of interdependent attributes. The procedure first deletes components associated with small latent roots of $\mathbf{X}'\mathbf{X}$ and then incorporates an analogue of the backward elimination procedure to eliminate the independent attributes, this deletion being based on minimal increases in residual sums of squares.

7.7.1.13 Case studies in the application of principal component analysis

Principal component analysis is probably the best known and most widely used ordination technique. The positions of the individuals can be plotted on pairs of rectangular Cartesian component axes. Such plots will show discontinuities if they exist in the data (e.g. see Blackith and Reyment, 1971), but it must be remembered that any such two-dimensional representation is distorted in that the other dimensions are not included. Gower and Ross (1969) showed that such distortions can be illustrated by superimposing the minimum spanning tree (see Section 7.9.1) of the points in the total dimensionality on to their representation in the reduced space. There has been much discussion about the use of principal component analysis in plant ecology, because of problems caused by the nature of the plot species distributions along environmental gradients. Too many papers have been written for detailed discussion here, but see, for example, Bray and Curtis (1957), Austin and Orlocci (1966), Beals (1973), Noy-Meir (1973), Whittaker (1973), Orlocci (1975).

Apart from Blackith and Reyment (1971), few collected examples of case studies in the application of principal component analysis exist, but examples of principal component analysis are given in more general books on multivariate analysis as a whole. One of the earliest of these books is that by Seal (1964) which is still worth reading as an introduction to multivariate techniques. Orlocci (1975) presents a wide range of multivariate techniques, including principal components, in vegetation research, and Mather (1976) does the same for physical geography. More recently, Gauch (1982) gives a range of applications in community ecology.

7.7.2 Factor analysis

7.7.2.1 Mathematical basis

In the principal component model, components were represented as linear additive functions of the original variables. In the model for factor analysis, however, it is assumed that each variate is represented as a linear additive function of K 'factors' common to all variates, together with a residual specific to each variate, i.e.:

$$x_j = \sum_{s=1}^{k} l_{js} f_s + e_j \qquad (j = 1, \ldots, p)$$

or $\mathbf{X} = \mathbf{LF} + \mathbf{E}$

where $\mathbf{X}$ and $\mathbf{E}$ are p-vectors, $\mathbf{F}$ is a k-vector, and $\mathbf{L}$ a $p + k$ matrix. The elements of $\mathbf{F}$ are called the *common factors*, while those of $\mathbf{E}$ are called *specific factors*. The elements of $\mathbf{L}$ are designated as the *loadings* of the factors.

Without loss of generality, it may be assumed that the common factors are uncorrelated with each other and with the specific factor, so that, with further assumptions of zero means and unit variances,

$$\mathbf{D} = \mathbf{L}\mathbf{L}' + \mathbf{V}$$

where $\mathbf{D}$ is the covariance matrix of the population of the p variate and $\mathbf{V}$ is the covariance matrix of the specific factors. Methods of estimating the loadings $\mathbf{L}$ are given by Lawley and Maxwell (1971), by the maximum likelihood solution is of particular interest.

The model

$$\mathbf{D} = \mathbf{L}\mathbf{L}' + \mathbf{V}$$

is assumed for the population of which the data matrix is a random sample, providing an estimated covariance matrix $\mathbf{C}$ of the population covariance matrix $\mathbf{D}$. If it can be assumed that the population has a multivariate normal distribution, then the sample covariance matrix $\mathbf{C}$ has a Wishart distribution with a known likelihood function whose maximum can be shown to be:

$$\mathbf{L}' - (\mathbf{V}\mathbf{L}'^{-1}(\mathbf{L}'\mathbf{V}^{-1}\mathbf{L} + \mathbf{I}))^{-1}\mathbf{C} = 0$$

As any orthogonal transformation $\mathbf{LH}$ of $\mathbf{L}$ will also satisfy the original model and

$$(\mathbf{LH})(\mathbf{LH})' = \mathbf{LHH}'\mathbf{L}' = \mathbf{LL}'$$

where $\mathbf{H}$ is orthogonal, a unique solution is obtained by solving for the values of $\mathbf{L}$ which make

$$\mathbf{L}'\mathbf{V}^{-1}\mathbf{L} = \mathbf{J} \qquad \text{(say)}$$

diagonal, and so

$$\mathbf{L}(\mathbf{I} + \mathbf{J}) - \mathbf{C}\mathbf{V}^{-1}\mathbf{L} = 0$$

is the basic equation for estimating $\mathbf{L}$.

Consequently,

$$\mathbf{LJ} = \mathbf{C}\mathbf{V}^{-1}\mathbf{L} - \mathbf{L}$$

so that the diagonal elements of $\mathbf{J}$ are the latent roots of:

$$\mathbf{V}^{-1/2}\mathbf{C}\mathbf{V}^{-1/2} - \mathbf{I}$$

with eigenvectors:

$$\mathbf{M} = \mathbf{V}^{-1/2}\mathbf{L}$$

and scaling given by:

$$\mathbf{M}'\mathbf{M} = \mathbf{M}'\mathbf{V}^{-1}\mathbf{L} = \mathbf{J}$$

The estimates of the loading $\mathbf{L}$ are given by $\mathbf{V}^{1/2}\mathbf{M}$.

In contrast to principal component analysis, factor analysis requires an assumption that the observations are taken from a population following a multivariate normal distribution — a feature which limits the usefulness of the model in data analysis.

It is also worth noting that the matrix $\mathbf{V}^{-1/2}\mathbf{C}\mathbf{V}^{1/2}$ is invariant to changes in scale of measurement, and is, in this respect, closely related to the correlation matrix $\mathbf{S}^{-1/2}\mathbf{C}\mathbf{S}^{1/2}$, where $\mathbf{S} = \text{diag}(\mathbf{C})$, but with the specific variances playing the same standardizing role that is played by the variances. In effect, there is a close relationship between factor analysis and principal component analysis of a correlation matrix, and it can be shown that the projected value of scaled co-ordinates $\mathbf{V}^{1/2}\mathbf{x}$ on to k dimensions after a principal component analysis of $\mathbf{V}^{-1/2}\mathbf{C}\mathbf{V}^{-1/2}$ differs from the estimates of factor scores only by a scale factor for each of the k axes. As a result, if the factor scores of an individual observation are regarded as its co-ordinates, similar configurations are obtained by both methods. If $\mathbf{V}$ is proportional to $\mathbf{S}$, a principal component analysis can be expected to give similar results to a factor analysis.

7.7.2.2 Practical considerations

The basic principle of factor analysis is that examination of the correlation matrix may show that a few of the attributes are highly correlated, while the remainder are not significantly correlated with them or with each other. If the purpose is to seek a basic pattern, we might consider retaining the attributes which show high inter-correlations and discarding the remainder. If the values of unity in the principal diagonal of the correlation matrix are replaced with the commonalities (i.e. the common factor variances), some of the eigenvalues, those associated chiefly with the unique variances, would become zero or nearly so. The search for commonalities is a major problem. It is necessary to decide in advance how many eigenvalues and eigenvectors we are interested in, and there is no way to be sure that the correct number has been chosen; it seems to be a matter of luck (Williams, 1976). The next step is to calculate the chosen number of eivenvalues and eigenvectors of the modified correlation matrix, and to normalize the eigenvectors so that the sum of squares of each vector equals the corresponding eigenvalue. The eigenvectors can then be used to provide an improved estimate of the commonalities, and the process is iterated until the commonality estimates are reasonably constant. In practice, this method is no longer used, more efficient methods being available.

7.7.2.3 Differences between factor analysis and principal component analysis

It is useful, at this stage, to emphasize the important differences between principal component analysis and factor analysis. Both methods have in common the aim to find a small number of hypothetical attributes (components or factors) which contain the essential information expressed in a larger number of observed attributes. Hence, the dimensionality of the data is reduced by utilizing interdependence. Factor analysis was originally developed for the analysis of scores obtained by individuals on batteries of psychological tests, and it still has many applications in that field, although its use has spread to other fields, notably geography (e.g. see Jöreskog *et al.*, 1976). Originally, the term *factor analysis* included principal component analysis, but it is important to distinguish between two methods which have similar mathematical models, but are not exact as representations of the data based on certain assumptions. The differences between the techniques are largely connected with differences in these assumptions.

The fundamental differences between principal component analysis and factor analysis lie in the ways in which the factors are defined, and in the assumptions about the nature of the residuals. In principal component analysis, the factors (components) are determined with a maximum variance constraint. In factor analysis, the factors are defined to account maximally for the inter-correlation of the attributes. In principal component analysis, the residual terms are assumed to be small, and a large part of the total variance of an attribute is assumed to be important. In factor analysis, there is no such assumption, and only that part of an attribute is used that participates in the correlation with other attributes. In both methods, the residuals are assumed to be uncorrelated with the factors. In principal component analysis, there is no assumption about correlations between the residuals, whereas the residuals in factor analysis are assumed to be uncorrelated.

7.7.2.4 Case studies

Blackith and Reyment (1971) suggest that it is very hard to discuss factor analysis without generating more heat than light; it is the most controversial of the multivariate methods. Factor analysis was proposed originally as a model for a well-defined problem in educational psychology, but it acquired a bad reputation among mathematicians and was largely ignored outside the field of psychology, where it still finds most of its applications. The method and criticisms were discussed by Cattell (1965), Blackith and Reyment (1971), and Marriott (1974) and several books have been written on this topic (e.g. Lawley and Maxwell, 1971).

Factor axes (and component axes) may be rotated to determinable positions in which they are not necessarily, or generally, orthogonal. Sneath and Sokal (1973) considered that this rotation makes scientific sense in that the factors underlying the covariation pattern of the characters in nature are themselves undoubtedly correlated, but they pointed out that there are practical problems. Clifford and Stephenson (1975) went so far as to state that 'It is likely that in the future factor analysis will play an increasingly important role in ecological studies.' On the other hand, Gower (1967b) considers it doubtful if factor analysis really is a helpful way of viewing biological data, and Blackith and Reyment (1971) ask: 'Could it not be that factor analysis has persisted precisely because, to a considerable extent, it allows the experimenter to impose his preconceived ideas on the raw data?'

7.7.3 Principal co-ordinate analysis

7.7.3.1 Background

Principal co-ordinate analysis is a computationally simple, but powerful, procedure which has wide applications. It arose from a dissatisfaction with many reported applications of factor analysis and principal component analysis in classification studies, particularly in the biological literature. If many, or all, of the attributes are qualitative, product-moment correlations between attributes may be inappropriate. In such a case, the analysis requires the use of a matrix of association coefficients, or some function of an association coefficient which is regarded as the distance between two objects. Some analysts applied the techniques of principal component analysis and fact that the standard underlying assumptions are not even approximately satisfied in such an analysis, obtained results which were successful, in that the expected relative magnitude of inter-object distances were recovered.

The problem was solved in a classic paper by Gower (1966). A simplified account is given by Gower (1967b). Suppose we have a data matrix $\mathbf{X}$ $(n \times p)$. Any row vector $\mathbf{x}_i'$ gives the co-ordinates of a point p_i which represents the position of the ith object in p-dimensional space. From $\mathbf{X}$ we can derive a $\mathbf{Q}$ matrix, of order n, of coefficients of association between objects. The eigenvectors of the $\mathbf{Q}$ matrix give the the co-ordinates of the points q_i which are the projections of points p_i on the new axes. The problem is to find the correct scaling for each eigenvector and the inter-object distance (p_i, p_j) when this scaling is used. Gower (1966) showed that, if the eigenvectors are normalized so that the sum of squares of the elements of a vector equals the corresponding eigenvalues, then these vectors define a Euclidean space (a 'Gower space'). The eigenvectors can be used as orthogonal axes, the vector elements being the co-ordinates of the objects in the new space. Given certain conditions, the Gower system is Euclidean even if the original measures were not.

7.7.3.2 Mathematical basis

Suppose we have a matrix $\mathbf{A}$, which is symmetric and of order n, the elements of which are some form of coefficient of association or distance between individuals. Matrix $\mathbf{A}$ has n eigenvalues and associated eigenvectors $\mathbf{c}_j$, which are the columns of a matrix $\mathbf{C}$. The elements of the ith row of $\mathbf{C}$ are taken as the co-ordinates of a point q_i in Euclidean n-space. The Pythagorean distance between two points q_i and q_j in this space is:

$$d_{ij} = \sum_{k=1}^{q} [(c_{ik} - c_{jj})^2]^{1/2}$$

Gower (1966) showed that, if the eigenvectors are normalized so that the sum of squares of a vector is equal to the corresponding eigenvalue, then:

$$d^2{}_{ij} = a_i + a_{jj} - 2a_{ij}$$

That is, the inter-object distances $\Delta(p_i, p_j)$ map into the Euclidean Gower space, regardless of whether or not the original distance properties are Euclidean, as long as those properties are suitable for representing the interrelationships between the individuals as expressed in the equation above. For example, if **A** is a similarity matrix or a formal product-moment correlation matrix between the individuals ($\Delta(p_i, p_j)$ will be zero for complete identification and will attain its maximum value for complete opposites. In both these cases, the diagonal elements of A are unity, so that:

$$[\Delta(P_i, P_j)]^2 = 2(1 - a_{ij})$$

It also follows from the equation $d_{ij} = a_{ii} + a_{jj} - 2a_{ij}$ that, if we put $a_{ij} = \frac{1}{2} d^2{}_{ij}$ and $a_{ii} = 0$, then $\Delta(p_i, \mathrm{p}_j) = d_{ij}$, and this gives a direct method of finding the co-ordinates of a set of points given their inter-point distances d_{ij}.

The validity of the above theory depends upon matrix **A** fulfilling two conditions. First, it must be symmetric; otherwise, the relationship between the equations does not hold. Fortunately, most of the commonly used association or similarity measures (described in detail by Sneath and Sokal, 1973) are symmetrical, although Williams *et al.* (1971) gave an example which failed in this respect. The second requirement is for **A** to be positive semi-definite; otherwise, one or more of the eigenvalues will be negative and part of the space will be imaginary, and, therefore, non-Euclidean. Fortunately, most commonly used measures appear to define positive semi-definite matrices, although missing values in the original data matrix can destroy this property.

Having set out the above theory for mapping the original measures into a Euclidean space, Gower (1966) then considered the use of principal component analysis on the matrix of co-ordinates of the points q_i, to find the best fit in fewer dimensions. He showed that this result can be achieved by transforming matrix **A** to matrix **B** such that:

$$b_{ij} = a_{ij} - r_i - c_j + g$$

where r_i is the mean of the ith row, c_j is the mean of the kth row, and g is the grand mean. The rows and columns of **B** sum to zero, and consequently **B** has a zero root. Matrix **B** must be positive semi-definite as before for **A**.

7.7.3.3 *Computational procedures*

The steps in the computation of the principal co-ordinate analysis are as follows.

(i) Form the association matrix **A**. Perhaps the most important aspect of this type of analysis is the choice of a suitable measure. This choice is discussed in more detail.
(ii) Transform **A** to **B** by the equation $b_{ij} = a_{ij} - r_i - c_j + g$.
(iii) Calculate the eigenvalues and eigenvectors of **B** and normalize each eigenvector so that its sum of squares is equal to its corresponding eigenvalue.

It is clear that it is very easy to run into practical computing problems in calculating the eigenvalues and eigenvectors of a matrix of order n. When n is large, both time and storage space can become a problem even on modern computers. One way of overcoming this problem is to tridiagonalize the matrix using an algorithm which needs only the lower half matrix to be stored (e.g. the TRED3 algorithm of Martin *et al.*, 1971). As only a few of the largest eigenvalues will be required, time can be saved by using an algorithm written especially for this purpose (e.g. the RATQR algorithm of Reinsch and Bauer, 1971). The corresponding eigenvectors of the tridiagonal matrix can then be calculated (e.g. by using the inverse iteration procedure in algorithm TRISTURM of Peters and Wilkinson, 1971). These eigenvectors will also need to be back-transformed to those of the original matrix (e.g. by algorithm TRBAK3 or Martin *et al.*, 1971). Another approach has been suggested by Lefkovitch (1976), and is based on the fact that, if **Y** is a transformation of the data matrix **X** such that $\mathbf{YY}'$ is a similarity or distance matrix, the principal co-ordinates of the n objects can be obtained from the eigenvalues and eigenvectors of $\mathbf{Y}'\mathbf{Y}'$. This solution is also possible with mixed data types.

7.7.3.4 *Interpretation*

Principal co-ordinate analysis has more limited uses than principal component and factor analysis methods. Its main aim is the graphical treatment of the data. As n points can be fitted into $n - 1$ dimensions, principal co-ordinate analysis must yield at least one zero eigenvalue. If the requirements mentioned above are satisfied, the r non-zero roots will be positive. The eigenvectors corresponding to the q largest positive eigenvalues give the co-ordinates of the points q_i, which are the projections of the points p_i on to the best-fitting subspace of dimensionality q. If an eigenvalue is small, then the contribution $(c_{ik} - c_{jk})^2$ to the distance between q_i and q_j will also be small. Thus, the only co-ordinates which contribute much to the distance are those with large eigenvalues which have wide variation in their vector elements. In many applications, it is found that the distances can be expressed adequately in terms of two or three such vectors. As in principal component analysis, the sum of squares of the residuals will be the difference between the trace of the matrix **B** and the sum of its q largest roots.

The first principal co-ordinate maximizes the total squared distance between the objects, the second maximizes the total squared distance in the space orthogonal to the first, and so on. The positions of the points can be plotted on pairs of principal co-ordinate axes, and the resulting ordination charge can be examined for pattern. The points are centred at the origin, because the columns of **C** sum to zero. In order to interpret each axis, we could calculate the correlation coefficient between each column of **C** and each of the original attributes.

Because

$$\mathbf{B} = \mathbf{CC}'$$

any diagonal element b_{ii} is the squared distance of point q_i from the origin O. Furthermore, an off-diagonal element b_{ij} is the cosine of the angle between the vectors from the centroid O to q_i and q_j times the product of distances Oq_i and Oq_j. The better the approximation given by the first q vectors, the closer q_i will be to p_i.

If matrix **A** consists of measures which are not real Euclidean distances, **B** may not be positive semi-definite, and therefore may have negative eigenvalues. Provided that these negative values are sufficiently close to zero, they can, in practice, be ignored.

Williams (1976) drew attention to an interesting application of principal co-ordinates in the analysis of factorial experiments. If the original treatments are orthogonal, the main effects tend to dispose themselves on separate eigenvectors. Furthermore, if the system has a very large error contribution, this may itself be patterned and may separate out as a distinct vector. Consequently, the true main effects can be tested against an error term from which an unwanted pattern has been partitioned out. Although this procedure appears to be both effective and powerful, it arouses misgivings in statisticians because the underlying rationale is not fully understood.

Another application also described by Williams (1976) is canonical co-ordinate analysis. This analysis can be applied to data in which two batteries of attributes have been measured on the same individuals. The two batteries of data are submitted to principal co-ordinate analysis separately, and a canonical correlation analysis is carried out on the two sets of principal co-ordinate vectors. Because of the orthogonality of the latter, the canonical correlation characteristic equation is considerably simplified; $\mathbf{B}_{11}$ becomes an identity matrix, as does $\mathbf{R}_{11}{}^{-1}$ and $\mathbf{R}_{22}{}^{-1}$. The equation thus reduces to:

$$|\mathbf{R}_{12}\mathbf{R}_{21} - \lambda^2\mathbf{I}| = 0$$

If only the principal co-ordinate vectors corresponding to a few of the larger eigenvalues have been used, the 'noise' itself will have been discarded.

7.7.3.5 *Alternative distance measures*

Any transformation of **A** will result in a transformation in the distance between q_i and q_j, with a consequent distortion in the configuration of the objects. This distortion may result in curvature effects on the ordination graphs, particularly if **A** has been formed from presence-absence data (Jöreskog *et al.*, 1976).

Pythagorean distance calculated from values of variates has nonsensical physical dimensions if the variates are measured on different scales. To overcome this difficulty, the variates can be normalized, usually by the sample standard deviation, although other normalizations could be used, for example the variate mean (when zero is not arbitrarily located), the range, or even the cube root of the sample third moment (Gower, 1966). The formula:

$$d^2{}_{ij} = \sum_{k=1}^{p} (x_{ik} - x_{jk})^2$$

makes no attempt to allow for correlations.

Boratynski and Davies (1971) investigated the taxonomic value of male coccids using 12 different numerical methods and 3 principal co-ordinate analyses, with 101 characters. They concluded, tentatively, that principal co-ordinate methods using non-parametric measures of association are, perhaps, best suited to the analysis of coded multistate data. They admitted, however, that principal co-ordinate analysis is unlikely to yield seriously misleading results.

7.7.3.6 *Duality with principal component analysis*

Two techniques are defined as being dual to one another when they both lead to a set of n points with the same inter-point distances. Gower (1966) showed that principal co-ordinate analysis, if applied to a matrix whose elements are $-\frac{1}{2}d^2{}_{ij}$, where d_{ij} is the Pythagorean distance between objects i and j calculated from the original variates, is dual to a principal component analysis of the sums of squares and products matrix of the variates. He gave two important special cases of this duality: (i) Sokal's measure of taxonomic distance, and (ii) the simple matching coefficient deviation.

If the principal components are derived from the correlation matrix, the inter-object distances are given by:

$$\Delta^2{}_{ij} = \sum_{k=1}^{p} \frac{(x_{ik} - x_{jk})^2}{s_k}$$

where s_k is the standard deviation of the kth variate. This distance is the measure of taxonomic distance proposed by Sokal and others (see Sneath and Sokal, 1973). A principal co-ordinate analysis of the matrix

$$\mathbf{\Delta}_{ij} = \sum_{k=1}^{p} \frac{x_{ij}x_{jk}}{s^2_k}$$

will lead to a reduced dimensional configuration identical to that obtained from a principal component analysis of the correlation matrix.

If **S** is the matrix of simple matching coefficients, then principal co-ordinate analysis of such a matrix will give points $[2(1 - s_{ij})]^{1/2}$ apart. The same points would be obtained by a principal component analysis of a matrix of corrected sums of squares and products. In fact, principal component analysis of (0, 1) data is exactly equivalent to assuming that the individuals are represented by points whose distance apart are proportional to $(1 - S_{ij})^{1/2}$.

7.7.3.7 Case studies

Factor or component scores from factor or principal component analysis can be plotted on pairs of rectangular Cartesian axes, and these scores provide a means of describing inter-object relationships. However, covariances or correlations are not the only bases, and may not necessarily be the best, on which to examine such relationships. The basis of principal co-ordinates is in a definition of inter-object similarity, or association, and the calculation of a similarity matrix, of order $n \times n$. Given such a matrix, the steps in the procedures are analogous to those of component analysis, but the interpretation of the results is different.

Analysis of inter-object similarity was first applied in geology by Imbrie and Purdy (1962). In this study, similarities were defined with respect to properties of constituents, for example mineral species, which make up a rock or sediment. Gower's method of principal co-ordinate analysis is based on the use of a distance or dissimilarity matrix. The ability to ordinate a set of objects given only their dissimilarities can be useful in ecological studies, and there are some circumstances in which a particular dissimilarity measure might be preferred. For example, one might wish to emphasize dominance and thus use the Bray–Curtis measure, or, perhaps, be more concerned with relative properties and use the Canberra metric (Clifford and Stephenson, 1975). Principal co-ordinate analysis is particularly useful when there are missing values or missing variates. In such a case, a correlation type of similarity measure is reasonably robust and reliable, whereas replacing the missing values by estimates or guesses is not usually satisfactory (Marriott, 1974).

Again, there are numerous descriptions of the use of the technique in ecology and the environmental sciences, but no texts devoted solely to case studies of its application.

7.7.4 Reciprocal averaging

Where some of the attributes to be used in a multivariate ordination model are quantitative, the use of principal component analysis, while apparently robust, has little theoretical justification. Principal co-ordinates analysis, on the other hand, can still often be used, even on wholly qualitative data, by devising indices of similarity which are then translated into distances by formulae such as:

$$d^2{}_{ii} = 1 - s_{ii}$$

where the similarity s_{ij} between individuals in the ratio of 'matches' to the number of attributes compared, a 'match' occurring when the attribute is either present or absent in both individuals.

However, one of the most valuable methods devised for the ordination of multivariate data which are essentially qualitative is that of reciprocal averaging described by Hill (1973). The model is especially appropriate for the analysis of presence–absence data or absence of species on sample quadrats. Geometrically, these data may be regarded as a set of points at the vertices of a hypercube, for which the ordination does not depend on the explicit use of the distances between the vertices.

Reciprocal averaging depends on a series of successive approximations during which the individuals are given an arbitrarily chosen set of starting scores, ideally chosen to represent some gradient suspect *a priori* as being reflected by the data. Average scores are then computed for attributes from which new, rescaled averages are calculated for the individuals. After a sufficient number of iterations, the final attribute scores converge to the same row vector, and the eigenvalue of the first axis is a measure of the extent to which the range of the scores contracts in one iteration.

When the first axis has been obtained, the second axis is considered, and a good starting point for the scores of this second axis may be obtained by using a set of scores which are close to the final scores for the first axis. Before iteration, however, these scores have to be adjusted by subtracting a multiple of the final first axis. A simple example, together with the appropriate computer algorithm, is given by Hill (1973).

The process is essentially a repeated cross-calibration which derives a unique, one-dimensional ordination of both the attributes and the individuals. It is called *reciprocal averaging* precisely because the attribute scores are averages of the individual scores, and, reciprocally, the individual scores are averages of the attribute scores. The final scores do not depend on the initial trial scores, but a good choice of inital scores considerably reduces the number of iterations required. The whole procedure is otherwise mathematically very similar to both principal component and principal co-ordinate analysis, and can be extended to cover quantitative as well as qualitative data.

Benzécri's (1973) method of *correspondence analysis*, based on the use of a contingency table for assessing similarities, shares many of the properties of reciprocal averaging, in that the similarity must be defined jointly and symmetrically. The method permits the simultaneous presentation of objects and attributes as points on the same pair of co-ordinate axes, so that mutual dependence can be interpreted. An extensive school of application of this method exists in the French ecological literature. An excellent introduction to the theory and application of correspondence analysis is given by Greenacre (1984).

7.8 DISCRIMINATION

The discrimination between two groups of individuals on the basis of measures of several attributes is a classical problem of which Fisher (1936) provided the earliest solution. He postulated a linear function of the measurements on each variable such that an individual can be assigned to one or other of the two groups with the least chance of being misclassified. Such a discriminant may be written as:

$$Z = a_1x_1 + a_2x_2 + \cdots + a_nx_n$$

where **a** is the vector of discriminant coefficients and **x** the vector of observations or measurements made on an individual which is to be assigned to one or other of the two groups. From this model, however, we are only considering the possibility of two groups, and, while we may decide that some individual cannot be assigned with any confidence to one of the two groups, we are not considering the formation of other groups. Thus, the descrimination assumes the imposition of an *a priori* grouping of the individuals, information which is external to the analysis itself. The group may be of very different sizes, i.e. with markedly different numbers of individuals. Again, we will seek to represent the P variables in as few dimensions as possible, but we will wish to emphasize the between-group variability at the expense of the within-group variability.

7.8.1 Mathematical basis

For any effective analysis of the problem, we must assume that the within-group variances and covariances are homogeneous, with **W** representing the pooled within-groups covariance matrix, and **B** the between-groups covariance matrix. In order to find linear combinations of the variables that are orthogonal and that successively maximize the between-group variance by comparison with the within-group variance, we maximize the criterion:

$$V = \frac{\mathbf{V}'\mathbf{B}\mathbf{I}}{\mathbf{I}'\mathbf{W}\mathbf{I}}$$

and this may be shown to be equivalent to solving:

$$(\mathbf{B} - \lambda\mathbf{W})\mathbf{I} = 0$$

where λ is the maximum of $\mathbf{V}$.

For consistency, we require that:

$$|\mathbf{B} - \lambda\mathbf{W}| = 0$$

leading once again to an equation of degree p having p solutions, $\lambda_1, \lambda_2, \ldots, \lambda_p$, with a vector $\mathbf{I}_j$ corresponding to each λ_j ($j = 1, 2, \ldots, p$) and giving the required linear combinations. Because the equation can be written as:

$$(\mathbf{W}^{-1}\mathbf{B} - \lambda\mathbf{I}')\mathbf{I} = 0$$

the λ_j are effectively the eigenvalues of $\mathbf{W}^{-1}\mathbf{B}$ with corresponding eigenvectors $\boldsymbol{\lambda}_j$, and, because the λ_j are the stationary values of:

$$\frac{\mathbf{I}'\mathbf{B}\mathbf{I}}{\mathbf{I}'\mathbf{W}\mathbf{I}}$$

the vector corresponding to the largest eigenvalue gives the direction along which the between-group variance is maximum relative to the within-group variance. The vector corresponding to the next largest eigenvalue gives the direction maximizing the remaining between-group variance relative to within-group variance, and so on.

The transformation of any vector $\mathbf{x}$ is thus given by $\mathbf{L}'\mathbf{X}$, and the space described by all such vectors is called the canonical variate space. Further, as $\mathbf{L}'\mathbf{W}\mathbf{L}$ is a diagonal matrix, the disadvantage of the arbitrary scaling of the λ_j may be overcome by normalizing as follows:

$$\mathbf{L}'\mathbf{W}\mathbf{L} = \mathbf{I}$$

The analysis represents the simultaneous reduction of $\mathbf{B}$ and $\mathbf{W}$ to diagonal form, in which the mean of the kth population in the original space is x_k, and this becomes $\mathbf{L}'\mathbf{x}_k$ in the canonical variate space. The distance between the kth and jth mean in the canonical variate space is:

$$(\mathbf{x}_k - \mathbf{x}_j)'\mathbf{L}\mathbf{L}'(\mathbf{x}_k - \mathbf{x}_j)$$

Canonical variate analysis is an extension of the idea of the discriminant function between two populations.

7.8.2 Computational aspects

Discrimination between two groups is characteristically simpler than discrimination between several groups, in that we need only seek a single discriminant function. Effective algorithms for the calculation are given by Blackith and Reyment (1971), and Davies (1971).

The extension of discriminant theory to more than two groups presents no great difficulties. All of the procedures for the two-group case generalize in a fairly obvious way. There are two possible ways of preceeding (Marriott, 1974). One possibility is to consider the groups in pairs, and estimate the discriminant function between each pair. An alternative is to consider the functions that maximize the variance ratio, or the variability between the groups considered together. These are the canonical variates although usually they are standardized on the matrix **W**, so that they have unit variance within groups. Both methods can be considered as generalizations of the discriminant function of two groups. The two approaches lead, in general, to slightly different answers, but only because the elimination of non-significant factors is carried out at different stages. The choice between the two is largely a matter of convenience.

In the first approach, the discriminant function between any two groups is calculated from the means and the pooled dispersion matrix **W**. If the distances between all pairs are significant, there are $\frac{1}{2}g(g-1)$ discriminant functions. The alternative method is to calculate the latent roots of $\mathbf{WS}^{-1}$ by canonical analysis. If the first k are significant, and the value of $L^* = (\lambda - \lambda_{k+1})\ldots(\lambda - \lambda_{g-1})$ is not significant, the first k canonical variates only are used. Original variates which are irrelevant may be discarded from these canonical variates which are then used to derive allocation rules as in the first method.

The two methods are not inconsistent. It is possible that a distance judged 'just significant' by the first will be judged 'not significant' by the second, and the discriminant functions will have slightly different coefficients unless all the canonical variates are used. Which approach is chosen depends essentially on whether the canonical variates give a worthwhile simplification. If $k = 1$, a single discriminant function is sufficient for all the groups, and the allocation rule is simply based on a dissection of the range of this function. If $k = 2$, a scatter diagram, with the two canonical variates as axes, can be drawn to show all the groups and the lines dividing them, based on discriminant functions which are linear functions of the two canonical variates. If $k = 3$, a three-dimensional model is needed to present the data. When $k > 3$, the advantage of using canonical variates is slight. If k is large, there may be some reduction of dimensionality, but it may still be better to consider doing several separate anaylses.

Algorithms for canonical variate analysis are, again, given in Blackith and Reyment (1971) and Davies (1971).

7.8.3 A distribution-free method

When the basic assumptions of the discriminant analysis are not justified for a particular set of data, the following technique has been suggested by Kendall and Stuart (1979).

(i) The range of each attribute should be divided into three non-overlapping parts:
 (a) one end containing members of group A only;
 (b) a middle, containing members of A and B;
 (c) the opposite end, containing members of group B only.
 If the extreme individuals belong to the same group, only (c) exists.

(ii) Select as x_2 the attribute with the largest number of individuals in (a_1) and (b_1).

(iii) Assign all observations in (a_1) to A, all in (b_1) to B. Record the innermost values of (a_1) and (b_1), and the first instruction in the allocation rule is to allot to A or B all individuals having values outside these limits.

(iv) Select as x_2 the attractive with the largest number of individuals of (c_1) in (a_2) and (b_2).

 The second instruction in the allocation rule then allots individuals unallocated by x_1 in accordance with the value of x_2.

(v) This process can be continued until either all individuals are disposed of, or until all the attributes have an individual of the same group at both ends of their range. Richards (1972) has pointed out that there is no reason why an attribute should not be used more than once in different steps of the rule.

This method depends only on the ranking of the separate attributes. It is, however, very inefficient when there are many attributes and none of them discriminates efficiently. The method could be modified by considering combinations of attributes, but such a modification would destroy some of the generality of the method. The real difficulty, however, is to know when to stop. It is necessary to decide whether the discrimination afforded by a particular attribute is real, or merely a chance effect. Some attributes will show differences that are obviously significant, but a decision on whether the ranking associated with the best of several measures — not, of course, independent — is or is not a chance effect requires the use of bootstrap or jackknife techniques (Efron, 1982).

7.8.4 Generalized distance

The concept of the *generalized distance* between two populations was suggested by Mahalanobis (1936) and its properties have been investigated chiefly by the Indian statisticians. The following account again follows closely that of Marriott (1974). Suppose $\mathbf{d}$ is the vector of differences between the x means in groups A and B ($d_i = x_{iA} - x_{iB}$; $i = 1, \ldots, p$). The statistic $D^2 = \mathbf{d}'\mathbf{V}^{-1}\mathbf{d}$ is the estimate of a corresponding parameter, dependent of the means of the two groups and the dispersion matrix within groups. This parameter is known as the squared generalized distance between the groups.

The generalized distance and its estimate have the following properties.

(i) They are scale-independent. The value of D^2 is unaltered if the x's are multiplied by arbitrary constants, or replaced by a set of linear combinations which are not linearly dependent (e.g. and set of principal components).

(ii) They take account of correlations between the variates.

(iii) The value of D is the difference between the mean values of the discriminant function, regarded as a linear function of the x's for the two groups, divided by its standard deviation.

The distribution of D^2 is known, and a significance test for difference between the two groups is given by:

$$F(p, n-g+1) = D^2(1/n_A + 1/b_B)^{-1}\frac{(n-p-1)}{p(n-g)}$$

where n is the total number of observations.

The test is generalized to the p-variate case of the ordinary t-test. It is sometimes expressed in terms of Hotelling's T. If

$$T^2 = \frac{D^2}{1/n_A + 1/n_B}$$

a significance test is given by

$$F(p, n-g-p+1) = T^2\frac{n-p-1}{p(n-g)}$$

When $p = 1$ and $g = 2$, this expression reduces to the ordinary t-test for the differences between the means of two groups.

The distribution of D^2 when the generalized distance in the population is not zero is based in the same way on the non-central F distribution (see Rao, 1965). The distribution has been used to compare distances based on different numbers of variates (Rao, 1950) or between corresponding groups in different populations.

7.8.5 Case studies

As before, while discriminant function analysis figures largely in most texts on multivariate analysis, there are relatively few case studies of the application of the technique in ecology, as opposed to taxonomy. Jeffers (1978a) gives one case study of the discrimination between sites on Signy Island, in the South Orkney Islands, with and without vascular plants. Norris and Barkham (1970) made a comparison of some Cotswold beechwoods by multiple discriminant analysis.

Hill (1977) has reported on the use of simple discriminant functions to classify quantitative phytosociological data, and Nielsen *et al.* (1973) have made a statistical evaluation of geobotanical and biogeochemical data by discriminant analysis. Discriminant analysis in tree nutrition research has also been reported by White and Mead (1971), and Valentine and Houston (1979) have used a discriminant function to identify mixed oak stand susceptibility to gypsy moth defoliation.

7.9 CLUSTER ANALYSIS

Classification involves the recognition of similarities between, and the grouping of, the individuals of the basic data matrix. A classification may have more than one purpose, but the paramount purpose is to describe the relationships of objects to each other, and to simplify the relationships so that general statements can be made about classes of individuals. An important distinction is between monothetic and polythetic classifications. *Monothetic* classifications are those in which the classes established differ by at least one property which is uniform within the members of each class. In *polythetic* classifications, the classes are groups of individuals or objects that share a large proportion of their attributes, but do not necessarily agree in any one attribute. A corollary of polythetic classification is the requirement that many attributes can be used to classify the individuals. However, once a classification has been established, only a few attributes are generally necessary to allocate individuals to the proper group. Classifications based on many attributes will be general. They are unlikely to be optimal for any single purpose, but might be useful for a great variety of purposes. By contrast, a classification based on few attributes might be optimal with respect to those attributes, but would be unlikely to be of general use (Sokal, 1974).

Hence, classification of a data set results in a reduction of the amount of information that is necessary to describe the data, but, if the classification is efficient, there is little or no reduction in the amount of information contained in the data. Furthermore, classifications that describe relationships between individuals from a defined population should generate hypotheses, possibly the main scientific justification for the exercise.

If dissection is to be carried out, the basis of the dissection must be clearly defined. For example, an ecologist may regard the vegetation as essentially continuously changing, but changing more rapidly in some regions than others. He will therefore wish to treat these zones of maximum gradient as if they were discontinuities, and so sharpen them by an appropriate technique (Williams, 1971). Some cluster methods have properties which make them useful for different types of dissection, for example the various minimum-variance methods and methods of the association analysis type. The flexible clustering strategy of Lance and Williams (1967) may also be useful for this purpose.

Cluster analysis of data representing the distribution of points in multi-dimensional space, where the distances between pairs of points are defined as some function of the observed sample values, has become a popular method of data analysis. The usual purpose of the analysis is to group the points in multi-dimensional space into (usually) disjoint sets which it is hoped will correspond to marked features of the sample. The grouped sets of points may themselves be grouped into larger sets, so that all the points are eventually classified hierarchically. This hierarchical classification can be represented diagrammatically in the form of a dendrogram showing the degree of relationship between individuals, and, ideally, a scale indicating the level of similarity between the suggested groupings. There are many forms of cluster analysis and classification, and the critical review of Cormack (1971) should be read by anyone intending to embark on the use of multivariate models for this purpose. In this Handbook, only a few of these forms will be considered.

7.9.1 Minimum spanning tree

Suppose that n points are given in two or more dimensions. A *tree* is defined as any set of straight-line segments joining pairs of points such that:

(i) no closed loops occur;
(ii) each point is visited by at least one line;
(iii) the tree is connected.

The length of the tree is the sum of the length of its segments, and, for any set of n points for which the lengths of all possible segments are known, it is possible to define a tree of minimum length which spans the points. Efficient algorithms for computing this minimum spanning tree are given by Gower and Ross (1969). The concept of the minimum spanning tree is of particular value in helping with the interpretation of diagrammatic representations of multi-variate data, and also as a first stage in the single linkage cluster analysis considered below.

7.9.2 Single linkage cluster analysis

This method of cluster analysis was proposed by Sneath (1957) as a helpful way of summarizing taxonomic relationships in the form of dendrograms, where the relationships are expressed in terms of taxonomic distances between every pair of samples, measured on some convenient scale. The method entails clustering the individual samples by comparing their distances with a series of increasing threshold distances, these threshold distances usually being increased by small constant steps rather than continuously. It can be shown that the method of clustering is closely related to the minimum spanning tree and the clusters at any level can be derived from the minimum spanning tree by deleting all segments of length greater than the defined level. Because some detail on the exact distances between samples may be lost when several links join two threshold levels, the dendrogram derived from single linkage cluster analysis may not be exactly the same as the dendrogram derived from the minimum spanning tree.

Single linkage cluster analysis has the disadvantage of producing long clusters of 'chained' samples under certain conditions, and these elongated clusters are generally regarded as being undesirable. On the other hand, evidence of a continuous sequence of intermediate samples can be informative. In addition, unlike most other methods of cluster analysis, single linkage gives exactly the same results by aggregating small clusters into larger clusters as by dividing larger clusters into smaller ones. This property enables the method to be used for much larger numbers of samples than many other techniques of cluster analysis, and makes it especially convenient for preliminary analysis of very large sets of multivariate data.

7.9.3 *k* means clustering

Wishart (1968) proposed an entirely different approach to numerical classification, and the following account of the method is given by Marriott (1974).

> If the criterion for the separation of groups is that there are several nodes in multivariate space, an obvious line of attack is to look for the nodes directly. Wishart did this by defining 'dense points' as the centres of hyperspheres of minimum radius containing a given number of points, and then expanding the hyperspheres to associate the other points with these centres. During this expansion, there is a continuous revision, the dense points moving to give minimum radius to the sphere associated with the number of points contained, and new dense points being defined as the radius increases. The process continues until all points have been classified, and, at this stage, they are divided into several groups giving the final classification. During the expansion, new group centres may emerge, and the groups already formed may be combined. If the distribution has only a single node, there may be only a single group with no discontinuity in the multivariate space.
>
> The process depends on one parameter, i.e. the number of points defining the original dense points or nodes. It is an agglomerative hierarchical procedure, and if this parameter $k = 1$, it reduces to ordinary single-link clustering. Nevertheless, when $k > 1$, it differs from other hierarchical techniques in its aims and conclusions. The purpose of the analysis is to find a natural grouping, and the intermediate steps are of no importance. It is theoretically possible to construct a dendrogram to represent the steps leading to the final grouping, but the intermediate stages consist of one or more groups and additional isolated points.
>
> The method is suitable for clustering both continuous variables and binary

attributes. It is not satisfactory for a mixture of the two types of attributes, and discrete or coarsely grouped attributes are apt to be troublesome. The definition of dense points avoids any assumption of an underlying distribution, and there is no sampling theory or significance test associated with the method. The aim is to detect clustering of the observations. If, in fact, the data are a sample from a distribution of some sort, it is not clear how effective the method is at rejecting spurious clustering due to sampling fluctuations, which will depend on the value chosen for k. Provided that k is not very small, it will not suggest a grouping when the data themselves give no indication of heterogeneity.

One defect of the method is that the definition of dense points in terms of spheres make it less effective when variates are highly correlated and the contours surrounding the modes are elongated ellipses. The difficulty could be overcome in the case of continuous variables by working with principal components or principal co-ordinates. Though it is technically a variation of agglomerative hierarchical clustering, the possibility of varying k gives Wishart's method far greater flexibility that other methods in this class. If the aim is to find a natural grouping, rather than to construct a dendrogram, it is effective and unlikely to give misleading results. On the whole, it is probably the best practical classification technique at present available.

In summary, Wishart's method has the following advantages:

(i) it is a direct approach designed to identify the nodes of the underlying distribution, or the clustering of the results;
(ii) it is unlikely to suggest a completely spurious grouping;
(iii) no sampling theory is invoked — though when the observations really are a sample, the properties of the method are not known.

Its weaknesses are:

(i) it is suitable only for continuous variables or binary attributes, but not both together;
(ii) for continuous variates, rather large samples may be needed;
(iii) it is sensitive in detecting elongated modes;
(iv) the choice of the value of k may affect the conclusions.

7.9.4 The use of components in cluster analysis

In some problems, we may wish to examine whether the n objects can be classified into groups, or clusters, so that the points within a cluster are close together, but the clusters themselves are, ideally, far apart. The problems involved in cluster analysis are discussed above, but here we can discuss briefly how principal component analysis can be of use. The discriminatory power of principal components can serve as a clustering technique of great generality (see examples in Blackith and Reyment, 1971). Plotting points on pairs of orthogonal component axes can help in several ways. First, it may suggest the suitability (or otherwise) of a particular form of analysis, for example if there are clearly defined and separate groups and whether these are spherical or elongated. Second, it may show why a particular technique has not given satisfactory results, and it may suggest alternatives. Finally, it may confirm that a suggested clustering looks reasonable and fits the observations realistically.

The orthogonality of the principal component transformation means that it is distance- and angle-preserving in an r-dimensional space. How well plots of points on pairs of orthogonal component axes represent the real configuration of the points depends on how well this confirmation is preserved in the reduced space. One way of looking for distortions in such plots is to superimpose on them the minimum spanning tree described above. It can also be useful to plot histograms of the frequency distributions of points along selected components. The procedure will show whether or not there are several nodes along a component (e.g. Webster and Burrough, 1972).

The investigator may have decided to accept the first q dimensions as preserving sufficient of the total variance for practical purposes. The configuration of the points in this q-dimensional space can be studied by calculating distances between pairs of points. It also might be worth computing the distances in r-dimensional space, and examining the distances between the differences in the two spaces to see if they are uniformly small (Rao, 1964).

Because the components are orthogonal, the distances can be simple Euclidean distances. As Euclidean distance depends on the scale of the attributes, it is unlikely to have much meaning if some attributes have a much greater range of values than others. Hence, it is generally used only when all the measurements have been standardized in some way (Marriott, 1974). Principal components are dimensionless, but, as calculated using the orthogonal eigenvectors, they have different variances, equal to the eigenvalues. Hence, if component values are to be used for the calculation of Pythagorean distances, it is preferable to normalize the components to unit sums of squares (Hope 1968). In effect, this is a question of the weighting given to each component in its contribution to the distance. By normalizing each component, it is effectively being weighted according to its variance.

There will be occasions, however, where this simple procedure of plotting multivariate data as a series of two-dimensional projections, with the minimum spanning tree to indicate the closeness of the point in multivariate space, will not be sufficient. When these occur, a simple transformation described by Andrews *et al.* (1971) may be used to obtain easily interpreted plots of high-dimensional data. This transformation embeds the data in an ever higher dimensioned, but easily visualized, space of functions, and then plots the functions.

For each point, expressed either as principal components or canonical variates, the function:

$$f_x(t) = x_1\sqrt{2} + x_2 \sin(t) + x_3 \cos(t) + x_4 \sin 2(t) + x_5 \cos(2t) + \cdots$$

(where $x_1, x_2, x_3, \ldots$ are the component values and t is variable $-\pi < t < \pi$) is plotted over the range $-\pi < t < \pi$. This function transforms the set of points into a set of lines drawn across the page in such a way that the mean of the functions of n observations corresponds to the mean of the observations themselves. The function representation also preserves distances, so that distance between two functions is proportional to Euclidean distance between the corresponding points. Even more important, the representation preserves variances, so that, if the components of the data are uncorrelated (as will usually be the case in our applications of the function) with a common variance σ^2, then the function value of t, $f_x(t)$, has a variance which is given by:

$$\text{var} \mid f_x(t) \mid = \sigma^2(\tfrac{1}{2} + \sin^2(t) + \cos^2(t) + \sin^2(2t) + \cos^2(2t) + \cdots).$$

The variability of the plotted function is almost constant across the graph, a fact which considerably simplifies this interpretation.

7.9.5 Other uses of components

If there is any reason to hypothesize that an object may belong to one of two or more groups which can be discriminated by a particular component, then the fit between such groups and the component values can be tested by analysis of variance (Hope, 1968). If the component values have been calculated using the orthonormal eigenvectors of a covariance or correlation matrix as described above, the total sum of squares of a component is $n - 1$ times its eigenvalue. The between-groups sum of squares can be calculated by squaring the mean component value of each group, multiplying the result by the number in the group, and then summing over all the groups. The within-groups sum of squares can be obtained by subtraction. For exploratory purposes, the more complicated designs in the analysis of variance may prove useful. If the hypothesis is that the groups may be discriminated by reference to two or more components, the multivariate analysis of variance may be useful (Rao, 1964), bearing in mind the need to satisfy certain requirements. It is worth bearing in mind the fact that, in practice, it may turn out that some of the components with smaller eigenvalues are better discriminators between groups with respect to the differences between means than are the first few components with the larger eigenvalues. Other uses are possible if the data can be referred to some sort of geographical grid, for example in stratified sampling (Patterson *et al.*, 1978).

Lefkovitch (1976) described a computationally simple divisive method for hierarchical clustering. He showed that the vectors of principal co-ordinates, in decreasing order of their eigenvalues, indicate the succession levels of the hierarchy of a dendrogram, and the signs of the vector elements indicate the group membership. Consider the following matrix **C**:

Object A	1.49	0.09	1	0
B	1.49	0.9	−1	0
C	−0.97	−1.13	0	0
D	−1.26	0.61	0	−0.5
E	−1.26	0.61	0	0.5
Eigenvalue	8.56	2.48	2	0.5

Starting from the left, the first division is into (A, B) and (C, D, E). At the next level, (C, D, E) divides into (C) and (D, E). After the following level, (A, B) divides, and at the final level (D, E) divides. Small differences in the numerical values of the co-ordinates do not alter the hierarchical relationships, which depend only on the signs.

There are three assumptions on this method: (i) $\mathbf{XX}'$ is an appropriate description of the inter-object relationships; (ii) the objects do, in fact, belong to disjoint, rather than overlapping, groups; (iii) a group formed by co-ordinates i and j is either unchanged by including co-ordinate $j+1$ or it is divided into two. Then the between-group squared distance is maximized by this method. Lefkovitch noted that computation can be simplified by transforming the data matrix **X** into **Y** and finding the eigenvalues and eigenvectors of $\mathbf{Y}'\mathbf{Y}$; the eigenvectors can then be converted to principal co-ordinates. This method can be used with a variety of attribute types.

7.9.6 Association analysis

Association analysis is a technique for obtaining a monothetic, divisive hierarchy for a set of data representing the recorded presences or absences of a number of attributes. The advantages of the technique (Williams, 1971) are as follows.

(i) Group definitions are simple and unambiguous. The groups are defined in terms of the presence or absence of the chosen attributes.
(ii) The majority of groups remain stable as additional entities are added to the data matrix. Alterations will occur, however, if a sufficient number of new entities affects the priorities in the choice of attributes.
(iii) Computation is relatively fast, because there is usually more interest in the upper than the lower level of the hierarchy, and the process can be halted at the required level.
(iv) At least in theory, divisive strategies begin classification when the total information available is a maximum.

Divisive methods are particularly suitable for handling large data matrices. They can also be employed in data elimination to reduce large matrics to a size where agglomerative methods can be employed. Two methods have been employed in the determination of the division attributes — these depending on information theory and those depending on χ^2. Although the χ^2 methods have now largely been superseded by information theory measures, some interest still remains in their use where it is not necessary to handle mixed data, i.e. presence or absence data mixed with quantitative variables (Lance and Williams, 1968a, b).

Given a set of binary attributes for a set of entities, the χ^2 are calculated for all attributes taken in pairs. These are then summed over all attributes and that with the largest $\Sigma_2^n \chi^2$ is used as the basis for dividing the set of entities into two subsets — those possessing the attribute and those lacking it. Each subset is further subdivided in the same manner as the original array of data until the required number of subsets is obtained, or until none of the χ^2 values exceed a pre-set probability level.

Algorithms for divisive χ^2 strategies based on binary data were produced by Williams and Lambert (1960) for vegetation analysis. Programs capable of handling mixed data and allowing for missing entries have been written by Lance and Williams (1968b). The technique is now regarded as having been superseded by indicator species analysis (see below), although it continues to have some useful properties.

7.9.7 Indicator species analysis

In ecological applications, the individuals of the basic data matrix are ordered by the first axis of a reciprocal averaging ordination, and the individuals are then divided into two groups at the centre of gravity of the ordination. Five 'indicator species' are chosen by the function:

$$I_j = | m_2/M_1 - m_2/M_2 |$$

where I_j is the indicator value of species j (assuming the value 1 if the species is a perfect indicator and a value 0 if it has no indicator value);
m_1 is the number of individuals in which species j occurs on the negative side of the dichotomy;
m_2 is the number of individuals in which species j occurs on the positive side of the dichotomy;
M_1 is the total number of individuals on the negative side of the dichotomy;
M_2 is the total number of individuals on the positive side of the dichotomy.

The five species with the highest indicator value are then used to construct an 'indicator score' for the whole set of individuals and to define an 'indicator threshold' which corresponds with the dichotomy.

The whole process may be repeated for the second and subsequent reciprocal averaging axes, so that the individuals are again divided by the same method, continuing as far as possible. No satisfactory rule for stopping the subdivision has so far been devised, so that there is a degree of arbitrariness both in the selection of thresholds and the number of subdivisions. The indicator scores can be regarded as providing an ordination of the individuals on a six-point scale, an admittedly crude ordination, but one which can be done quickly. The aim is to mirror the original reciprocal averaging ordination sufficiently closely for the analysis to the used as a criterion for classifying the individuals.

An algorithm for the calculations of the indicator species analysis is given by Hill (1973).

7.9.8 Case studies

The literature on classification and cluster analysis in ecology is vast. For an introduction, it is probably best to start from the general texts of Orlocci (1975), Greig-Smith (1983), Gauch (1982), Williams (1976), and Blackith and Reyment (1971).

7.10 FITTING RELATIONSHIPS BETWEEN GROUPS OF VARIABLES

7.10.1 Mathematical basis

As an alternative to the assumption of an *a priori* structure imposed upon the individuals of the basic matrix, it may be assumed that the attributes of the matrix are divided into two sets, with r and q attributes in each set, so that $p = r + q$. This is equivalent to writing the data matrix:

$$\mathbf{X} = [\mathbf{X}_1 \quad \mathbf{X}_2]$$

Where $\mathbf{X}_1$ has n rows and p columns and $\mathbf{X}_2$ has n rows and q columns.

The covariance matrix computed from the basic matrix may be partitioned as:

$$\mathbf{S} = \begin{bmatrix} \mathrm{p}\mathbf{A}\mathrm{p} & \mathrm{p}\mathbf{C}\mathrm{q} \\ \mathrm{q}\mathbf{C}'\mathrm{p} & \mathrm{q}\mathbf{B}\mathrm{q} \end{bmatrix}$$

From this matrix, it will frequently be of interest to find the linear combinations:

$$\mathbf{u}_i = \mathbf{l}'_i\mathbf{X}_1$$
$$\mathbf{v}_i = \mathbf{m}'_i\mathbf{X}_2$$

where $i = 1, 2, 3, \ldots, s$, with the property that the correlation of $\mathbf{u}_1$ and $\mathbf{v}_1$ is greatest, the correlation of $\mathbf{u}_2$ and $\mathbf{v}_2$ is greatest among all linear combinations uncorrelated with $\mathbf{u}_1$ and $\mathbf{v}_1$, and so on for all possible pairs.

The correlation between any two linear combinations:

$$\mathbf{u} = \mathbf{l}'\mathbf{X}_1$$
$$\mathbf{v} = \mathbf{m}'\mathbf{X}_2$$

is given by:

$$p^2 = \frac{(\mathbf{l}'\mathbf{Cm})^2}{(\mathbf{l}'\mathbf{Al})(\mathbf{m}'\mathbf{Bm})}$$

and this is the criterion which we seek to maximize. It may be shown that this maximum is equivalent to the solution of the equations:

$$(\mathbf{C}'\mathbf{A}^{-1}\mathbf{C} - p^2\mathbf{B})\mathbf{m} = 0$$
$$(\mathbf{C}\mathbf{B}^{-1}\mathbf{C}' - p^2\mathbf{A})\mathbf{l} = 0$$

where p^2 is the stationary value.

The two matrices:

$$\mathbf{C}'\mathbf{A}^{-1}\mathbf{C} - \mathbf{B}$$
$$\mathbf{C}\mathbf{B}^{-1}\mathbf{C}' - \mathbf{A}$$

have identical roots, and the eigenvectors corresponding to these roots give the coefficients for the linear combinations. Just as with canonical variates:

$$\mathbf{L}'\mathbf{AL} = \mathbf{D}_1$$
$$\mathbf{M}'\mathbf{BM} = \mathbf{D}_2$$
$$\mathbf{L}'\mathbf{CM} = \mathbf{D}_3$$

where $\mathbf{D}_i$ is a diagonal matrix. If $q < p$ and all vectors of type 1 are in the matrix $\mathbf{L}(q \times q)$, then

$$\mathbf{M} = \mathbf{A}^{-1}\mathbf{CL}$$

and $\mathbf{M}$ is $(p \times q)$, where the remaining $p - q$ columns of $\mathbf{M}$ correspond to zero canonical correlations.

7.10.2 Multiple regression

The theory behind multiple regression is so well known that is is hardly necessary to give an account of the technique in this Handbook. Excellent texts already exist, notably Snedecor and Cochran (1967), Williams (1959) and Sprent (1969). Just about every computer installation has a multiple regression package, although many of them are computationally suspect.

7.10.3 The use of components in regression analysis

A serious problem in regression analysis is collinearity. The estimators of the coefficients depend on the inverse of the covariance matrix of the regressors. If one variable is a linear function of another, the coefficients in a regression equation which includes them are indeterminate, for the determinant of the covariance matrix vanishes. If some of the eigenvalues of the covariance matrix are small, its determinant is also small, and the coefficients will be ill-determined (Kendell, 1975).

Because the principal components of the matrix of regressor variables are orthogonal, it is natural to consider using them as regressors. Massey (1965) discussed principal component regression analysis in some detail. In an empirical study, he found that the principal component regression technique gave larger values for R^2 while employing fewer regressors than did their classical counterparts in three out of four cases examined. He concluded that the methods involved are useful because: (i) they permit rapid calculation of the correlations between a dependent variable and each of the components, and (ii) they refer the regression results back to the projections of the original independent variables in the space spanned by the components included in a given regression. This partially overcomes the problem of identifying the components in order to give meaning to the regression coefficients. Jolliffe's (1982) warning about the need to include components with small eigenvalues in such calculations should, however, be heeded carefully.

Daling and Tamura (1970) used components, not as new variables, but as the reference frame to identify a near-orthogonal subset of explanatory variables. Selection of such variables minimizes overlapping of information supplied by explanatory variables in the regression. They suggested applying varimax rotation to a type D eigenvector matrix. The varimax criterion produces a matrix of vectors in each of which a few variables tend to have high loadings, while the rest have small or zero loadings. The regression of the dependent variable on each varimax factor is then calculated to identify the explanatory variables which, having minimum interdependence among themselves, appear to make a significant contribution to the variance of the dependent variable. Hawkins (1973) described an interactive method which has the advantage of enabling good alternative subsets of predictors to be found easily. This method is based on a varimax rotation of a type D eigenvector matrix, the rotated matrix being used to suggest possible variables.

More recently, Page and Fabian (1978) have used principal component regression to examine the relationships between disease and air pollution. Jolliffe (1972, 1973) discussed eight methods, four of which use principal components. He applied five of the eight techniques to real data including a multiple linear regression analysis. Mansfield *et al.* (1977) presented a method for reducing the number of inter-dependent attributes. The procedure first deletes components associated with small latent roots of $\mathbf{X}'\mathbf{X}$ and then incorporates an analogue of the backward elimination procedure to eliminate the independent attributes, this deletion being based on minimal increases in residual sums of squares.

7.10.4 Canonical correlations

The generalization of which multiple regression is a special case assumes that there are two sets of attributes $x_1, \ldots, x_p$ and $y_1, \ldots, y_q$, one set at least being random variables. If significance tests are to be used, it will also be assumed that this is jointly normally distributed about a mean dependent on the other set.

Given the dispersion matrix of the complete set of $p + q$ attributes, the correlation of any given linear combination of the x's with a given linear combination of the y's can readily be calculated. One of the possible pairs of linear combinations must have the maximum correlation, and this is called the first canonical correlation. The corresponding pair of linear combinations of the x's and y's are called the first canonical variables. Because they have an arbitrary scale factor and an arbitrary mean, these constants may be chosen so that each canonical variable has zero mean and unit variance.

The second canonical correlation and variable may then be defined by the second pair of variables, uncorrelated with the first pair, that have the maximum correlation. Similarly, the process may be continued until there are p pairs of canonical variables and p canonical correlations, assuming that $p < q$.

The resulting p pairs of variables have the following properties:

(i) all the correlations between them are zero, except those between the corresponding pairs;
(ii) the correlations between the corresponding pairs form a decreasing sequence;
(iii) each variable is standardized, i.e. has zero mean unit variance.

Any interpretation of the relationships between the two sets of attributes is now based on these canonical variables. Under certain rather strict assumptions, it is possible to test whether there is evidence of any relationship, whether the relationship is accounted for by the first pair, or first few pairs, of variables, and whether some of the original attributes can be omitted without significantly altering the conclusions. If reification of the canonical variables is possible, the nature of the relationship between the two sets of attributes may be clarified.

An algorithm for canonical correlation analysis is given by Anderson (1958). The only difficulty with the computation is the need for an efficient procedure for finding the eigenvalues and eigenvectors of a non-symmetric matrix Martin *et al.*, 1971; Bowdler *et al.*, 1971).

7.10.5 Case studies

There are very few examples of canonical correlation analysis being applied in ecology. Blackith and Reyment (1971) describe one or two such applications. Gillins (1979) also describes ecological applications of canonical analysis.

Acknowledgement

The help of Peter Howard in the writing of this chapter is gratefully acknowledged.

References

Anderberg, M. R. (1973). *Cluster analysis for applications*. New York, London: Academic Press.

Anderson, T. W. (1958). *An introduction to multivariate statistical analysis*. New York, Chichester: John Wiley.

Anderson, T. W. (1963). Asymptotic theory for principal component analysis. *Annals of Mathematical Statistics*, **34**, 122–128.

Anderson, T. W., and Goodman, L. A. (1957). Statistical inference about Markov chains. *Annals of Mathematical Statistics*, **28**, 89–110.

Andrews, D. F. (1972). Plots of high dimensional data. *Biometrics*, **28**, 125–136.

Anthony, T. F., Taylor, B. W. (1977). Analyzing the predictive capabilities of Markovian analysis for air pollution level variations. *J. Environ. Manage.*, **2**, 139–149.

Austin, M. P., and Noy-Meir, I. (1971). The problem of non-linearity in ordination: experiments with two-gradient models. *J. Ecol.*, **59**, 217–227.

Austin, M. P., and Orlocci, L. (1966). Geometric models in ecology II. An evaluation of some ordination techniques. *J. Ecol.*, **54**, 217–227.

Barkham, J. P. (1968). The ecology of the ground flora of some Cotswold beechwoods. University of Birmingham, PhD thesis.

Barraclough, R. M., and Blackith, R. E. (1962). Morphometric relationships in the genus *Ditylenchus*. *Nematologica*, **8**, 51–58.

Bartholomew, D. J., and Forbes, A. F. (1979). *Statistical techniques for manpower planning*. New York, Chichester: John Wiley.

Bartlett, M. S. (1950). Tests of significance in factor analysis. *British J. Psych. (Stat. Sect.)*, **3**, 77–85.

Beals, E. W. (1973). Ordination: mathematical elegance and ecological naivete. *J. Ecol.*, **61**, 23–35.

Bellman, R. (1960). *Dynamic programming*. Princeton: Princeton University Press.

Benzécri, J.-P. (1973). *L'Analyse des donnees*. Paris: Dunod.

Blackith, R. E., and Reyment, R. A. (1971). *Multivariate morphometrics*. London, New York: Academic Press.

Bhattacharya, R. N., *et al.* (1976). A Markovian stochastic basis for the transport of water through unsaturated soil. *J. Amer. Soc. Soil Science*, **40**, 465–467.

Binkley, C. S. (1980). Is succession in hardwood forests a stationary Markov process? *Forest Science*, **26**, 566–570.

Boratynski, K., and Davies, R. G., (1971). The taxonomic value of male *Coccoidea* (Homoptera) with an evaluation of some numerical techniques. *Bot. J. Linn. Soc.*, **3**, 57–102.

Botkin, D. B. (1977). Life and death in a forest: the computer as an aid to understanding. In: *Ecosystem modelling in theory and practice* (edited by C. A. S. Hall and J. W. Day), pp. 213–233. New York, Chichester: John Wiley.

Botkin, D. B., Janak, J. F., and Wallis, J. R. (1972a). Rationale, limitations and assumptions of a northeastern forest growth simulation. *IBM J. Res. Dev.*, **16**, 101–116.

Botkin, D. B., Janak, J. F., and Wallis, J. R. (1972b). Some ecological consequences of a computer model of forest growth. *J. Ecol.*, **60**, 849–872.

Bowdler, H., Martin, R. S., Reinsch, C., and Wilkinson, J. H. (1971). The QR and QL algorithms for symmetric matrices. In: *Handbook for automatic computation*; Vol. II, *Linear algebra* (edited by J. H. Wilkinson and C. Reinsch), pp. 227–250. Berlin, Heidelberg, New York: Springer-Verlag.

Bray, J. R., and Curtis, J. T. (1957). An ordination of the upland forest communities of southern Wisconsin. *Ecol. Monogr.*, **27**, 325–349.

Brockington, N. R. (1979). *Computer modelling in agriculture.* Oxford: Oxford University Press.

Buongiorno, J. and Michie, B. R., (1980). A matrix model of uneven-aged forest management. *Forest Science*, **26**, 609–625.

Cassell, R. F. and Moser, J. W., (1974). A programmed Markov model for predicting diameter distribution and species composition in uneven-aged forests. *Purdue Univ. Agric. Exp. Stn. Res. Bull.* **915**.

Cattell, R. B. (1965). Factor analysis: an introduction to essentials I. the purpose and underlying models. *Biometrics*, **21**, 190–215.

Clifford, H. T., and Stephenson, W. (1975). *An introduction to numerical classification.* London: Academic Press.

Colinvaux, P. A. (1973). *Introduction to ecology.* New York, Chichester: John Wiley.

Connell, J., and Slatyer, R. O. (1977). Mechanisms of succession in natural communities and their role in community stability and organization. *Am. Nat.*, **111**, 1119–1144.

Conway, G. R., and Murdie, G. (1972). Population models as a basis for pest control. In: *Mathematical models in ecology* (edited by J. N. R. Jeffers), pp. 195–214. Oxford: Blackwell Scientific.

Cooke, D. (In press). The description of plant succession data using a Markov chain model of plant-by-plant replacement.

Cooley, W. W., and Lohnes, P. R. (1971). *Multivariate data analysis.* New York, Chichester: John Wiley.

Cormack, R. M. (1971). A review of classification. *J. R. Statist. Soc., A*, **134**, 321–367.

Cunia, T., and Chevrou, R. B. (1969). Sampling with partial replacement on three or more occasions. *Forest Science*, **15**, 205–224.

Dale, M. B. (1975). On objectives of methods of ordination. *Vegetation*, **30**, 15–32.

Daling, J. R., and Tamura, H. (1970). Use of orthogonal factors for selection of variables in a regression equation — an illustration. *Appl. Statist.*, **19**, 260–268.

Dempster, J. P. (1975). *Animal population ecology.* London, New York, San Francisco: Academic Press.

Davies, R. G. (1971). *Computer programming in quantitative biology.* London, New York: Academic Press.

Debussche, M., Godron, M., Lepart, J. and Romane, F. (1977). An account of the use of a transition matrix. *Agro-Ecosystems*, **3**, 81–92.

Dent, J. B., and Blackie, M. J. (1979). *Systems simulation in agriculture.* London: Applied Science Publishers.

De Witt, C. T., and Goudriaan, J. (1974). *Simulation of ecological processes.* Wageningen, The Netherlands: Center for Agricultural Publishing and Documentation.

Drury, W. H., and Nisbet, I. C. T. (1973). Succession. *J. Arnold Abor.*, **54**, 331–368.

Dubois, D. M. (1979). State-of-the-art on predator – prey systems modelling. In: *State-of-the-art in ecological modelling* (edited by S. E. Jorgensen), pp. 163–217. Oxford: Pergamon.

Efron, B. (1982). *The jackknife, the bootstrap and other resampling plans.* Philadelphia, Pa.: Society for Industrial and Applied Mathematics.

Elger, F. E. (1954). Vegetation science concepts. I. Initial floristic composition — a factor in old field vegetation development. *Vegetation*, **4**, 412–417.

Fisher, R. A. (1936). The use of multiple measures in taxonomic problems. *Ann. Eugen.*, **8**a, 376–386.

Gabriel, K. R. (1971). The bipolot graphic display of matrices with application to principal component analysis. *Biometrics*, **58**, 453–467.

Gauch, H. G. (1982). *Multivariate analysis in community ecology.* Cambridge: Cambridge University Press.

Glaz, J. (1979). Probabilities and moments for absorption in finite homogeneous birth-death processes. *Biometrics*, **35**, 813–816.

Goodall, D. W. (1972). Building and testing ecosystem models. In: *Mathematical models in ecology*, (edited by J. N. R. Jeffers), pp 173–194. Oxford: Blackwell Scientific.

Gower, J. C. (1967). Multivariate analysis and multidimensional geometry. *Statistician*, **17**, 13–28.

Gower, J. C. (1971). A general coefficient of similarity and some of its properties. *Biometrics*, **27**, 857–871.

Gower, J. C. and Ross, G. J. S. (1969), Minimum spanning trees and single linkage cluster analysis. *Applied Statistics*, **18**, 54–64.

Greenacre, M. J. (1984). *Theory and applications of correspondence analysis.* London, New York: Academic Press.

Greig-Smith, P. (1983). *Quantitative plant ecology.* Oxford: Blackwell Scientific.

Hahn, H. H. and Eppler, B. (1979). Models of rivers. In: *State-of-the-art in ecological modelling*, (edited by S. E. Jorgensen), pp. 13–58. Oxford: Pergamon Press.

Hall, C. A. S., and Day, J. W. (1977). Systems and models: terms and basic principles. In: *Ecosystem modelling in theory and practice* (edited by C. A. S. Hall and J. W. Day), pp. 6–36. New York, Chichester: Wiley.

Hall, C. A. S., Day, J. W., and Odum, H. T. (1977). A circuit language for energy and matter. In: *Ecosystem modelling in theory and practice* (edited by C. A. S. Hall and J. W. Day), pp. 37–48. New York, Chichester: Wiley.

Harbaugh, J. W. and Bonham-Carter, G. (1970). *Computer simulation in geology.* London, Chichester. John Wiley & Sons.

Hawkins, D. M. (1973). On the investigation of alternative regressions by principal component analysis. *Appl. Statist.*, **22**, 275–286.

Hill, M. O. (1973). Reciprocal averaging: an eigenvector method of ordination. *J. Ecol.*, **61**, 237–251.

Hill, M. O. (1977). Use of simple discrimination functions to classify quantitative phytosociological data. *Proc. 1st Int. Symp. Data Analysis and Informatics, Versailles, September*, Vol. 1, pp. 181–196.

Holland, D. A. (1969). Component analysis: an aid to the interpretation of data. *Exp. Agric.*, **4**, 151–164.

Hope, K. (1968). *Methods of multivariate analysis.* London: University of London Press.

Horn, H. S. (1974). The ecology of secondary succession. *Ann, Rev. Ecol. Syst.*, **5**, 25–37.

Horn, H. S. (1975). Markovian properties of forest succession. In: *Ecology and evolution of communities* (edited by M. L. Cody and J. M. Diamond), pp. 126–211. Cambridge, Mass.: Harvard University Press.

Howarth, R. J., and Murray, J. W. (1969). The foraminiferida of Christchurch Harbour, England: A reappraisal using multivariate techniques. *J. Paleont.*, **43**, 660–675.

Imbrie, J., and Purdy, E. G. (1962). Classification of modern Bahamian carbonate sediments. *Amer. Ass. Petrol. Geol. Mem.*, **7**, pp. 253–272.

Innis, G. S., and O'Neill, R. V. (eds) (1979). *Systems analysis of ecosystems.* Maryland: International Co-operative Publishing House.

Ivimey-Cook, R. B., and Proctor, M. C. F. (1967). Factor analysis of data from an east Devon heath: a comparison of principal component and rotated solutions. *J. Ecol.*, **55**, 405–415.

James, A. T. (1977). Tests for a prescribed subspace of principal components. In: *Multivariate analysis IV* (edited by P. R. Krishnaiah), pp. 73–77. Amsterdam: North-Holland Publishing Co.

Jardine, N., and Sibson, R. (1971). *Mathematical taxonomy.* New York, Chichester: John Wiley.

Jeffers, J. N. R. (1965). Principal component analysis in taxonomic research. *For. Comm. Statist. Section paper no. 83*, 21 pp.

Jeffers, J. N. R. (Ed.) (1972). *Mathematical models in ecology.* Oxford: Blackwell Scientific.

Jeffers, J. N. R. (1978a). *An introduction to systems analysis; with ecological applications.* London: Edward Arnold.

Jeffers, J. N. R. (1978b). Design of experiments. *Stat. Checkl., Inst. Terr. Ecol.*, no. 1.

Jeffers, J. N. R. (1979). Sampling. *Stat. Checkl., Inst. Terr. Ecol.*, no. 2.

Jolicoeur, P., and Mosimann, J. E. (1960). Size and shape variation in the painted turtle, a principle component analysis. *Growth*, **24**, 339–354.

Jolliffe, I. T. (1972). Discarding variables in a principal component analysis. I. Artificial data. *Appl. Stat.*, **21**, 160–173.

Jolliffe, I. T. (1973). Discarding variables in a principal component analysis. II. Real data. *Appl. Stat.*, **22**, 21–31.

Jolliffe, I. T. (1982). A note on the use of principal components in regression. *Appl. Stat.*, **31**, 300–303.

Jöreskog, K. G., Klovan, J. E., and Reyment, R. A. (1976). *Geological factor analysis.* Amsterdam: Elsevier.

Jorgensen, S. E. (Ed.) (1979a). *State-of-the-art in ecological modelling.* Oxford: Pergamon Press.

Jorgensen, S. E. (1979b). State-of-the-art in eutrophication models.. In: *State-of-the-art in ecological modelling* (edited by S. E. Jorgensen). Oxford. Pergamon Press.

Kaiser, H. F. (1960).The application of electronic computers to factor analysis. *Educ. Psychol. Meas.*, **20**, 141–151.

Kendall, M. G. (1975). *Multivariate analysis.* London: Griffin.

Kendall, M. G., and Stuart, A. (1979). *The advanced theory of statistics*, Vol. 2 (4th edn). London: Griffin.

Kleijnen, J. P. C. (1975). *Statistical techniques in simulation.* New York: Dekker.

Krishnaiah, P. R., and Lee, J. C. (1977). Inference on the eigenvalues of the covariance matrices of real and complex multivariate normal populations. In: *Multivariate analysis IV* (edited by P. R. Krishnaiah), pp. 95–103. Amsterdam: North-Holland Publishing Co.

Krumbein, W. C. (1967). FORTRAN IV computer programs for Markov chain experiments in geology. *Computer Contribution, 13*, Kansas Geological Survey.

Krzanowski, W. J. (1971). The algebraic basis of classical multivariate methods. *The Statistician*, **20**, 51–61.

Lance, G. N., and Williams, W. T. (1967). A general theory of classification sorting strategies. I. Hierarchical systems. *Comp. J.*, **9**, 373–380.

Lance, G. N., and Williams, W. T. (1968a). Mixed-data classificatory programs. II. Devisive systems. *Aust. Comput. J.*, **1**, 82-85.

Lance, G. N., and Williams, W. T. (1968b). Note on a new information-statistic classificatory program. *Comp. J.*, **11**, 195.

Lassiter, R. R. (1979). Microcosms as ecosystems for testing ecological models. In: *State-of-the-art in ecological modelling* (edited by S. E. Jorgensen), pp, 127–161. Oxford: Pergamon.

Lassiter, R. R., Baughman, G. L. and Burns, L. A. (1979). Fate of toxic substances in the aquatic environment. In: *State-of-the-art in ecological modelling*, (edited by S. E. Jorgensen), pp. 219–246. Oxford, Pergamon Press.

Lawley, D. N. and Maxwell, A. E. (1971). *Factor analysis as a statistical method.* London: Butterworth.

Lefkovitch, L. P. (1976). Hierarchical clustering from principal co-ordinates. *Math. Biosci.*, **31**, 157–174.

Lembersky, M. R. and Johnson, K. N. (1975). Optimal policies for managed stands: an infinite horizon Markov decision process approach. *Forest Science*, **21**, 109–122.

Leslie, P. H. (1945). On the use of matrices in certain population mathematics. *Biometrika*, **35**, 183–212.

Leslie, P. H. (1948). Some further notes on the use of matrices in population mathematics. *Biometrika*, **35**, 213–245.

Lloyd, E. H. (1977). Reservoirs with seasonally varying Markovian inflows and their first passage times. *International Institute of Applied Systems Analysis*, RR-77-4.

McCammon, R. B. (1966). Principal component analysis and its application in large-scale correlation studies. *J. Geol.*, **74**, 721–733.

McCammon, R. B. (1968). Multiple component analysis and its application in classification of environments. *Bull. Amer. Ass. Petrol. Geol.*, **52**, 2178–2196.

McNeil, D. R. (1977). *Interactive data analysis.* Chichester, New York: John Wiley.

Maguire, B. J. (1979). Modelling of ecological process and ecosystems with particular response structures: a review and a new paradigm for diagnosis of emergent ecosystem dynamics and patterns. In: *State-of-the-art in ecological modelling* (edited by S. E. Jorgensen), pp. 59–126. Oxford: Pergamon.

Mahalanobis, P. C. (1936). On the generalised distances in statistics. *Proc natn. Inst. Sci. India*, **2**, 49–55.

Mansfield, E. R., Webster, J. T., and Gunst, R. T. (1977). An analytic variable selection technique for principal component regression. *Appl. Stat.*, **26**, 34–40.

Marriott, F. H. C. (1974). *The interpretation of multiple observations.* London, New York. San Francisco: Academic Press.

Martin, R. S., Reinsch, C., and Wilkinson, J. H. (1971). Householder's tridiagonalization of a symmetric matrix. In: *handbook for automatic computation*, Vol. II *Linear algebra*, (edited by J. H. Wilkinson and C. Reinsch), pp. 212–226. Berlin, Heidelberg, New York: Springer Verlag.

Massey, W. F. (1965). Principal components regression in exploratory statistical research. *J. Amer. Stat. Assn.*, **60**, 234–256.

Mather, P. M. (1976). *Computational methods of multivariate analysis in physical geography.* London, Chichester: John Wiley.

Maynard-Smith, J. (1974). *Models in ecology.* Cambridge: Cambridge University Press.

Milner, C., and Hughes, R. E. (1968). Methods for the measurement of primary productivity of grassland. *IBP Handbook No. 6*. Oxford: Blackwell Scientific.

Morrison, D. F. (1967). *Multivariate statistical methods*. New York, London, Sydney: McGraw-Hill.

Mosteller, F., and Tukey, J. W. (1977). *Data analysis and regression*. London: Addison-Wesley Publishing Co.

Munn, R. E. (Ed.) (1975). *Environmental impact assessment: principles and procedures. SCOPE Report No. 5*.

Newbould, P. J. (1967). Methods for estimating the primary production of forests. *IBP Handbook No. 2*. Oxford: Blackwell Scientific.

Nielsen, J. S., Brooks, R. R., Boswell, C. R., and Marshall, N. J. (1973). Statistical evaluation of geobotanical and biogeochemical data by discriminant analysis. *J. Appl. Ecol.*, **10**, 251–258.

Norris, J. M. (1971). Functional relationships in the interpretation of principal component analysis. *Area*, **3**, 217–220.

Norris, J. M., and Barkham, J. P. (1979). A comparison of some Cotswold beechwoods using multiple discriminant analysis. *J. Ecol.*, **58**, 603–619.

Noy-Meir, I. (1973). Data transformations in ecological ordinations I. Some advantages on non-centering. *J. Ecol.*, **61**, 329–341.

Noy-Meir, I., and Austin, M. P. (1970). Principal component ordination and simulated vegetational data. *Ecology*, **51**, 551–552.

O'Neill, R. V. (1971). Systems approaches to the study of forest floor arthropods. In: *Systems analysis and simulation in ecology*, Vol. I (edited by B. C. Patten), pp. 441–477. New York: Academic Press.

O'Neill, R. V., Ferguson, N., and Watts, J. A. (Eds) (1977). *A bibliography of mathematical modelling in ecology*. EDFB/IBP-75/5. Washington, D. C.: US Govt Printing Office.

Orlocci, L. (1975). *Multivariate analysis in vegetation research*. The Hague: Dr W. Junk.

Ortega, J. (1967). The Givens–Householder method for symmetric matrices. In: *Mathematical methods for digital computers*, Vol. II (edited by A. Ralston and H. Wilf), pp. 84–115. Chichester, New York, Sydney: John Wiley.

Ott, W. R. (Ed.) (1976). *Environmental modelling and simulation*. Virginia: National Technical Information Services.

Overton, W. S. (1977). A strategy of model construction. In: *Ecosystem modelling in theory and practice* (edited by C. A. S. Hall and J. W. Day), pp. 49–73. New York, Chichester: Wiley.

Page, W. P., and Fabian, R. G. (1978). Factor analysis: an exploratory methodology and management technique for the economies of air pollution. *J. environ. Management*, **6**, 185–192.

Patten, B. C. (Ed.) (1971). *Systems analysis and simulation in ecology*, Vol. I. New York: Academic Press.

Patten, B. C. (Ed.) (1972). *Systems analysis and simulation in ecology*, Vol. II. New York: Academic Press.

Patten, B. C. (Ed.) (1975). *Systems analysis and simulation in ecology*, Vol. III. New York: Academic Press.

Patten, B. C. (Ed.) (1976). *Systems analysis and simulation in ecology*, Vol. IV. New York: Academic Press.

Patterson, J. G., Goodchild, N. A., and Boyd, W. J. R. (1978). Classifying environments for sampling purposes using a principal component analysis of climatic data. *Agric. Meteorol.*, **19**, 349–362.

Peden, L. M., Williams, J. S. and Frayer, W. E. (1973). A Markov model for stand projection. *Forest Science*, **19**, 303–314.

Pelto, C. R. (1954). Mapping of multicomponent systems. *J. Geol.*, **62**, 501–511.

Peters, G. and Wilkinson, J. H. (1971). The calculation of specified eigenvectors by inverse iteration. In: *Handbook for automatic computation. Vol. II, Linear Algebra* (edited by J. H. Wilkinson and C. Reinsch), pp. 418–439. Heidelberg: Springer-Verlag.

Phillipson, J. B. (1966). *Ecological energetics*. London: Edward Arnold.

Pielou, E. C. (1969). *An introduction to mathematical ecology*. New York, Chichester: John Wiley.

Plinston, D. T. (1972). Parameter sensitivity and interdependence in hydrological models. In: *Mathematical models in ecology* (edited by J. N. R. Jeffers), pp. 237–248. Oxford: Blackwell Scientific.

Radford, P. J. (1972). The simulation language as an aid to ecological modelling. In: *Mathematical models in ecology* (edited by J. N. R. Jeffers), pp. 277–296. Oxford: Blackwell Scientific.

Raiffi, H. (1968). *Decision analysis — introductory lectures on choices under uncertainty*. Reading, Mass.: Adddison-Wesley.

Rao, C. R. (1950). Statistical inference applied to classificatory problems. *Sankhya*, **10**, 229–256.

Rao, C. R. (1964). The use and interpretation of principal component analysis in applied research. *Sankhya, Ser. A,* **26**, 329–358.

Rao, C. R. (1965). *Linear statistical inference and its applications*. Chichester, New York: John Wiley.

Rao, C. R. and Kshirsagar, A. M. (1978). A semi-Markovian model for predator-prey interactions. *Biometrics*, **34**, 611–620.

Reinsch, C. and Bauer, F. L. (1971). Rational QL transformation with Newton shift for symmetric tridiagonal matrices. In: *Handbook for automatic computation. Vol. II, Linear algebra* (edited by J. H. Wilkinson and C. Reinsch), pp. 257–265. Heidelberg: Springer-Verlag.

Richards, L. E. (1972). Refinement and extension of distribution-free discriminant analysis. *Applied Statistics*, **21**, 174–176.

Roberts, N. *et al.* (1983). *Computer simulation*. London: Addison-Wesley Publishing Company.

Ross, C. J. S. (1972). Stochastic model fittings by evolutionary operation. In: *Mathematical models in ecology* (edited by J. N. R. Jeffers), pp. 297–308. Oxford: Blackwell Scientific.

Rowen, H. S. (1976). Policy analysis as heuristic aid: the design of means, ends and institutions. In: *When values conflict* (edited by L. H. Tribe, C. S. Schelling and J. Voss). Cambridge, Mass.: Ballinger.

Schatzoff, M., and Tillman, C. C. (1975). Design of experiments in simulator validation. *IBM J. Res. Dev.*, **19**, 252–262.

Seal, H. (1964). *Multivariate statistical analysis for biologists*. London: Methuen and Co.

Searle, S. R. (1966). *Matrix algebra for the biological sciences*. New York, Chichester: John Wiley.

Shoemaker, C. A. (1977a). Pest management models of crop ecosystems. In: *Ecosystem modelling in theory and practice* (edited by C. A. S. Hall and J. W. Day), pp. 545–574. New York, Chichester: John Wiley.

Shoemaker, C. A. (1977b). Mathematical construction of ecological models. In:. *Ecosystem modelling in theory and practice* (edited by C. A. S. Hall and J. W. Day), pp. 75–114. New York, Chichester: Wiley.

Shugart, H. H., and O'Neill, R. V. (1979). *Systems ecology*. Stroudsburg, Pa.: Dowden, Hutchinson Ross.

Shugart, H. H., Crow, T. R., and Hett, J. M. (1973). Forest succession models: a rationale and methodology for modelling forest succession over large regions. *Forest Sci.*, **19** 203–212.

Simmonds, J. L. (1963). Application of characteristic vector analysis to photographic and optical response data. *J. Optical Soc. Amer.*, **53**, 986–974.

Skellam, J. G. (1972). Some philosophical aspects of mathematical modelling in empirical science with special reference to ecology. In: *Mathematical models in ecology* (edited by J. N. R. Jeffers), pp. 13–28. Oxford: Blackwell Scientific.

Skogerboe, G. V., Walker, W. R. and Evans, R. G. (1979). Modeling process for assessing water quality problems and developing appropriate solutions in irrigated agriculture. In: *State-of-the-art in ecological modeling*, (edited by S. E. Jorgensen), pp. 269–292. Oxford: Pergamon Press.

Slatyer, R. O. (1977). Dynamic changes in terrestrial ecosystems: paterns of change, techniques for study and application to management. *UNESCO MAB Technical Notes 4*.

Smeach, S. C. and Jernigan, R. W. (1977). Further aspects of a Markovian sampling policy for water quality monitoring. *Biometrics*, **33**, 41–46.

Smith, F. E. (1970). Analysis of ecosystems. In: *Analysis of temperate forest ecosystems* (edited by D. Reichle), pp. 7–18. New York: Springer-Verlag.

Sneath, P. H. A. (1957). Computers in taxonomy. *J. gen Microbiology*, **17**, 201–226.

Sneath, P. H. A., and Sokal, R. R. (1973). *Numerical taxonomy, the principles and practice of numerical classification*. San Francisco: W. H. Freeman.

Snedecor, G. W., and Cochran, W. G. (1967). *Statistical methods* (6th edn). Ames: Iowa State Univ. Press.,

Sokal, R. R. (1974). Classification: purposes, principles, progress, prospects. *Science*, **185**, 1115–1123.

Sokal, R. R. and Rohlf, F. J. (1962). The comparison of dendrograms by objective methods. *Taxon*, **11**, 33–40.

Sparks, D. N., and Todd, A. D. (1973). A comparison of FORTRAN subroutines for calculating latent roots and vectors. *Appl. Stat.*, **22**, 220–225.

Sprent, P. (1969). *Models in regression and related topics*, London: Methuen.

Stephens, G. R., and Waggoner, P. E. (1970). The forests anticipated from 40 years on natural transitions in mixed hardwoods. *Bull. Conn. Agric. Exp. Stn*, **77**, 1.58.

Tansley, A. G. (1935). The use and abuse of vegetational concepts and terms. *Ecology*, **16**, 284–307.

Tavare, S. (1979). A note on finite homogeneous continuous-time Markov chains. *Biometrics*, **35**, 831–834.

Taylor, C. C. (1977). Principal component and factor analysis. In: *The analysis of survey data*, Vol. 1, *Exploring data structures* (edited by C. A. O'Muircheartaigh and C. Payne), pp. 89–124. New York, Chichester: John Wiley.

Tukey, J. W. (1977). *Exploratory data analysis*. London: Addison-Wesley Publishing Co.

Udvardy, M. D. F. (1975). A classification of the biographical provinces of the world. *IUCN Occas. Pap. No. 18*.

UNESCO (1972). Expert panel in the role of systems analysis and modelling approaches in the programme on Man and the Biosphere (MAB). *UNESCO (MAB) Rep. Ser. No. 2*.

Usher, M. B. (1972). Developments in the Leslie matrix model. In: *Mathematical models in ecology* (edited by J. N. R. Jeffers), pp. 29–60. Oxford: Blackwell Scientific.

Usher, M. B. (1979). Markovian approaches to ecological succession. *Journal of Animal ecology*, **48**, 413–426.

Usher, M. B. and Parr, T. W. (1977). Are there successional changes in arthropod decomposer communities? *Journal of Environmental Management*, **5**, 151–160.

Valentine, H. T., and Houston, D. R. (1979). A discriminant function for identifying mixed-oak stand susceptibility to gypsy moth defoliation. *Forest Sci.*, **25**, 468–474.

Vandeveer, L. R. and Drummond, H. E. (1978). The use of Markov processes in estimating land use change. *Tech. Bull. No. 148*. Oklahoma: Agricultural Experimental Station.

Volterra, V. (1926). Variazione e fluttuazini del numero d'individui in specie animali conviventi. *Mem. Accad. Nazionale Lincei* (6) **2**, 31–113.

Waggoner, P. E. and Stevens, G. R. (1970). Transitional probabilities for a forest. *Nature*, **225**, 1160–1161.

Ware, K. D., and Cunia, T. (1962). Continuous forest inventory with partial replacement of samples. *Forest Sci. Monogr. No. 3*.

Webster, R., and Burrough, P. A. (1972). Computer-based soil mapping of small areas from sample data. I. Multivariate classification and ordination. *J. Soil Sci.*, **23**, 210–221.

White, E. H., and Mead, D. J. (1971). Discriminant analysis in tree nutrition research. *Forest Sci.*, **17**, 425–427.

Whittaker, J. H. (Ed.) (1973). *Ordination and classification of communities*. The Hague: Dr W. Junk.

Whittaker, R. H. (1967). Gradient analysis of vegetation. *Biol. Rev.*, **49**, 207–264.

Wilkinson, J. H. (1965). *The algebraic eigenvalue problem*. London: Oxford Univ. Press.

Williams, E. J. (1959). *Regression analysis*. London, New York, Chichester: John Wiley.

Williams, W. T. (1971). Principles of clustering. *Ann. Rev. Ecol. Syst.*, **2**, 303–326.

Williams, W. T. (Ed.) (1976). *Pattern analysis in agricultural science*. Amsterdam, Oxford, New York: Elsevier.

Williams, W. T., and Lambert, J. M. (1960). Multivariate methods in plant ecology. II. The use of an electronic digital computer for association analysis. *J. Ecol.*, **48**, 689–710.

Williams, W. T., Lance, G. N., Tracey, J. G. and Connell, J. H. (1969). Studies in the numerical analysis of complex rain-forest communities. IV. A method for the elucidation of small scale forest pattern. *J. Ecol.*, **57**, 635–654.

Williams, W. T., Clifford, H. T., and Lance, G. N. (1971). Group-size dependence: rationale for choice between numerical classifications. *Comp. J.*, **14**, 157–162.

Williams, W. T., Lance, G. N., Webb, L. J., and Tracey, J. G. (1973). Studies in the numerical analysis of complex rain-forest communities. VI. Models for the classification of quantitative data. *J. Ecol.*, **61**, 47–70.

Williamson, M. H. (1972). *The analysis of biological populations*. London: Arnold and Sons.

Wishart, D. (1968). *A FORTRAN II program for numerical classification*. St. Andrews University.

Appendix

BASIC PROGRAMS FOR SOME KEY ALGORITHMS

This Appendix contains computer programs in the BASIC language for some of the key algorithms mentioned in the main text, as follows:

DIFFEU	Solving differential equations using the Euler method.
DIFFEQ	Solving differential equations using the midpoint rule.
DIFFER	Solving differential equations using the Runge–Kutta approximation.
PREDAT	Plotting the trajectory of predator–prey populations.
MARK1	Computing the Markov transition matrix.
MARK2	Testing transition frequencies for the Markov property.
MARK3	Test of the stationarity of a Markov chain.
MARK4	Computing successive Markov probabilities.
AMARK	Calculating the properties of an absorbing Markov chain.
EMARK	Calculating the properties of an ergodic Markov chain.

DIFFEU: SOLVING DIFFERENTIAL EQUATIONS USING THE EULER METHOD

```
0010 REM ** SOLVE DIFFERENTIAL EQUATION USING EULER METHOD
0011 REM ** WRITTEN BY J.N.R.JEFFERS, INSTITUTE OF TERRESTRIAL ECOLOGY
0012 REM ** ORIGINAL PROGRAM FEBRUARY 1980
0013 REM
0014 REM ** PROGRAM COMPUTES THE VALUES FOR A DEFINED FUNCTION REPRESENTING
0015 REM ** A DIFFERENTIAL EQUATION IN X AND Y, FOR A GIVEN STEP SIZE BETWEEN
0016 REM ** A START AND END VALUE OF X, WITH A GIVEN STARTING VALUE OF Y
0017 REM
0018 REM ** DEFINE FUNCTION IN TERMS OF X AND Y
0019 REM
0020         DEF FNF(X,Y)=0.1*Y
0024 REM
0025 REM ** INPUT START AND END TIMES, STARTING VALUE AND STEP SIZE
0026 REM
0030         DISP "START TIME";
0040         INPUT A
0050         DISP "END TIME";
0060         INPUT Z
```

```
0070         DISP "STARTING VALUE";
0080         INPUT Y
0090         DISP "TIME STEP";
0100         INPUT H
0104 REM
0105 REM ** PRINT INITIAL VALUES
0106 REM
0110         PRINT A,Y
0114 REM
0115 REM ** SUBROUTINE TO PRINT SUCCESSIVE VALUES
0116 REM
0120         FOR X=A TO Z-H STEP H
0130             LET D=FNF(X,Y)
0140             LET Y=Y+D*H
0150             PRINT X+H,Y
0160         NEXT X
0165 REM
0170      END

END OF LISTING
```

DIFFEQ: SOLVING DIFFERENTIAL EQUATIONS USING THE MIDPOINT RULE

```
0010 REM ** SOLVE DIFFERENTIAL EQUATION USING MIDPOINT RULE
0011 REM ** WRITTEN BY J.N.R.JEFFERS, INSTITUTE OF TERRESTRIAL ECOLOGY
0012 REM ** ORIGINAL PROGRAM WRITTEN JANUARY 1980, ANNOTATED JANUARY 1982
0013 REM
0014 REM ** PROGRAM COMPUTES THE VALUES OF A DEFINED FUNCTION REPRESENTING
0015 REM ** A DIFFERENTIAL EQUATION IN X AND Y, FOR A GIVEN STEP SIZE BETWEEN
0016 REM ** A START AND END VALUE OF X, WITH A GIVEN STARTING VALUE OF Y.
0017 REM ** THE MIDPOINT RULE IS USED IN CALCULATING VALUES
0018 REM
0019 REM ** DEFINE FUNCTION IN TERMS OF X AND Y
0020 REM
0021         DEF FNF(X,Y)=0.1*Y
0024 REM
0025 REM ** INPUT START AND END TIMES, STARTING VALUE AND STEP SIZE
0026 REM
0030         DISP "START TIME";
0040         INPUT A
0050         DISP "END TIME";
0060         INPUT Z
0070         DISP "STARTING VALUE";
0080         INPUT Y
0090         DISP "TIME STEP";
0100         INPUT H
0104 REM
0105 REM ** PRINT INITIAL VALUES
0106 REM
0110         PRINT A,Y
```

```
0114 REM
0115 REM ** SUBROUTINE TO PRINT SUCCESSIVE VALUES
0116 REM
0120         FOR X=A TO Z-H STEP H
0130             LET D1=FNF(X,Y)
0140             LET D2=FNF(X+H/2,Y+D1*H/2)
0150             LET Y=Y+D2*H
0160             PRINT X+H,Y
0170         NEXT X
0175 REM
0180      END

END OF LISTING
```

DIFFER: SOLVING DIFFERENTIAL EQUATIONS USING THE RUNGE–KUTTA APPROXIMATION

```
0010 REM ** SOLVE DIFFERENTIAL EQUATION USING RUNGE-KUTTA
0011 REM ** WRITTEN BY J.N.R.JEFFERS, INSTITUTE OF TERRESTRIAL ECOLOGY
0012 REM ** ORIGINAL PROGRAM WRITTEN JANUARY 1980, ANNOTATED JANUARY 1982
0013 REM
0014 REM ** PROGRAM COMPUTES THE VALUES OF A DEFINED FUNCTION REPRESENTING
0015 REM ** A DIFFERENTIAL EQUATION IN X AND Y, FOR A GIVEN STEP SIZE BETWEEN
0016 REM ** A START AND END VALUE OF X, WITH A GIVEN STARTING VALUE OF Y.
0017 REM ** THE RUNGE-KUTTA SOLUTION IS USED IN CALCULATING VALUES
0018 REM
0019 REM ** DEFINE FUNCTION IN TERMS OF X AND Y
0020 REM
0021         DEF FNF(X,Y)=0.1*Y
0024 REM
0025 REM ** INPUT START AND END TIMES, STARTING VALUE AND STEP SIZE
0026 REM
0030         DISP "START TIME";
0040         INPUT A
0050         DISP "END TIME";
0060         INPUT Z
0070         DISP "STARTING VALUE";
0080         INPUT Y
0090         DISP "TIME STEP";
0100         INPUT H
0104 REM
0105 REM ** PRINT INITIAL VALUES
0106 REM
0110         PRINT A,Y
0114 REM
0115 REM ** RUNGE-KUTTA SUBROUTINE TO PRINT SUCCESSIVE VALUES
0116 REM
0120         FOR X=A TO Z-H STEP H
0130             LET T4=A+X*H
0140             LET T3=T4-H/2
```

```
0150            LET T2=T4-H
0160            LET K1=FNF(T2,Y)
0170            LET Z3=Y+K1*H/2
0180            LET K2=FNF(T3,Z3)
0190            LET Z3=Y+K2*H/2
0200            LET K3=FNF(T3,Z3)
0210            LET Z3=Y+K3*H
0220            LET K4=FNF(T4,Z3)
0230            LET Y=Y+(K1+2*K2+2*K3+K4)*H/6
0240            PRINT X+H,Y
0250         NEXT X
0255 REM
0260      END

END OF LISTING
```

PREDAT: PLOTTING THE TRAJECTORY OF PREDATOR–PREY POPULATIONS

```
0010 REM PROGRAM TO PLOT TRAJECTORY OF PREDATOR-PREY POPULATIONS
0020 DEF FNX(X)=4*X-0.25*X*X-2*X*Y
0030 DEF FNY(Y)=X*Y-3*Y
0040 ERASE
0050 INIMAGE DIS,2
0060 FRAME 8,5.6
0070 DISP "LIMITS OF X-AXIS";
0080 INPUT X8,X9
0090 DISP "LIMITS OF Y-AXIS";
0100 INPUT Y8,Y9
0110 SCALE X8,X9,Y8,Y9
0120 XAXIS 0,1
0130 YAXIS 0,1
0140 DISP "TIME INTERVAL";
0150 INPUT T8
0160 DISP "TOTAL TIME";
0170 INPUT T9
0180 DISP "NO OF PLOTTED POINTS";
0190 INPUT P
0200 DISP "STARTING VALUES-X,Y";
0210 INPUT X,Y
0220 IF X<0 THEN 330
0230 IF Y<0 THEN 330
0240 MOVE X,Y
0250 FOR I=1 TO P STEP 1
0260 FOR T=0 TO T9/P STEP T8
0270 LET X=X+FNX(X)*T8
0280 LET Y=Y+FNY(Y)*T8
0290 NEXT T
0300 PLOT X,Y
0310 NEXT I
```

```
0320 GOTO 200
0330 END

END OF LISTING
```

MARK1: COMPUTING THE MARKOV TRANSITION MATRIX

```
0010 REM ** PROGRAM TO COMPUTE MARKOV TRANSITION MATRIX
0020 REM ** FROM A SEQUENCE OF N INTEGERS
0021 REM ** WRITTEN BY J.N.R.JEFFERS, INSTITUTE OF TERRESTRIAL ECOLOGY
0022 REM ** PROGRAM WRITTEN AND ANNOTATED JANUARY 1982
0023 REM
0024 REM ** PROGRAM COMPUTES THE FREQUENCIES OF THE TRANSITIONS FROM ONE
0025 REM ** STATE TO ANOTHER FROM A SEQUENCE OF INTEGERS REPRESENTING THE
0026 REM ** SUCCESSIVE STATES. FREQUENCIES AND TRANSITION PROBABILITIES
0027 REM ** ARE PRINTED FOR SUBSEQUENT ANALYSIS.
0028 REM
0029 REM ** DIMENSION STATEMENT DEPENDS ONLY ON NUMBER OF STATES
0030 REM
0031        DIM P(20,20)
0034 REM
0035 REM ** INPUT NO OF STATES AND NO OF INTEGERS
0036 REM
0040        DISP "NO OF STATES";
0050        INPUT M
0060        DISP "NO OF DATA VALUES";
0070        INPUT N
0074 REM
0075 REM ** CLEAR WORKING MATRIX
0076 REM
0080        FOR I=1 TO M STEP 1
0090            FOR J=1 TO M STEP 1
0100                LET P(I,J)=0
0110            NEXT J
0120        NEXT I
0124 REM
0125 REM ** INPUT SUCCESSIVE INTEGERS AND COMPUTE FREQUENCIES
0126 REM
0130        INPUT I
0140        FOR K=2 TO N STEP 1
0150            INPUT J
0160            IF J<1 THEN 150
0170            IF J>M THEN 150
0180            LET P(I,J)=P(I,J)+1
0190            LET I=J
0200        NEXT K
0204 REM
0205 REM ** PRINT OBSERVED FREQUENCIES AND COMPUTE PROBABILITIES
0206 REM
```

```
0210          PRINT "OBSERVED FREQUENCIES"
0220          PRINT
0230          PRINT
0240          PRINT
0245 REM
0250          FOR I=1 TO M STEP 1
0260              LET S=0
0270              FOR J=1 TO M STEP 1
0280                  LET S=S+P(I,J)
0290                  PRINT P(I,J),
0300              NEXT J
0310              FOR J=1 TO M STEP 1
0320                LET P(I,J)=P(I,J)/S
0330              NEXT J
0340              PRINT
0350          NEXT I
0354 REM
0355 REM ** PRINT TRANSITION PROBABILITIES
0356 REM
0360          PRINT
0370          PRINT
0380          PRINT
0390          PRINT "TRANSITION PROBABILITIES"
0400          PRINT
0410          PRINT
0420          PRINT
0425 REM
0430          FOR I=1 TO M STEP 1
0440              FOR J=1 TO M STEP 1
0450                  PRINT P(I,J),
0460              NEXT J
0470              PRINT
0480          NEXT I
0485 REM
0490       END

END OF LISTING
```

MARK2: TESTING TRANSITION FREQUENCIES FOR THE MARKOV PROPERTY

```
0005 REM ** PROGRAM TO TEST TRANSITION FREQUENCIES FOR THE MARKOV PROPERTY
0006 REM ** WRITTEN BY J.N.R.JEFFERS, INSTITUTE OF TERRESTRIAL ECOLOGY
0007 REM ** WRITTEN AND ANNOTATED JANUARY 1982
0008 REM
0009 REM ** THE PROGRAM READS A MATRIX OF TRANSITION FREQUENCIES HELD AS
0010 REM ** DATA AND COMPUTES A CHI-SQUARE TEST THAT THE TRANSITIONS ARE
0011 REM ** FROM AN INDEPENDENT EVENTS PROCESS. REJECTION OF THE NULL
0012 REM ** HYPOTHESIS INDICATES THE ALTERNATIVE HYPOTHESIS THAT THE
0013 REM ** TRANSITIONS HAVE THE MARKOV PROPERTY.
```

```
0014 REM
0018 REM ** DIMENSION STATEMENT DEPENDS ON NO OF STATES
0019 REM
0020        DIM T(6,6),P(6,6),C(6)
0024 REM
0025 REM ** INPUT NO OF STATES
0026 REM
0030        DISP "NO OF STATES";
0040        INPUT M
0044 REM
0045 REM ** READ FREQUENCIES FROM DATA STATEMENTS AND ZERO ROW ACCUMULATORS
0046 REM
0050        FOR I=1 TO M STEP 1
0060            FOR J=1 TO M STEP 1
0070                READ T(I,J)
0080            NEXT J
0090            LET C(I)=0
0100        NEXT I
0104 REM
0105 REM ** COMPUTE TRANSITION PROBABILITIES AND ROW TOTALS
0106 REM
0110        LET Z=0
0120        FOR I=1 TO M STEP 1
0130            LET S=0
0140            FOR J=1 TO M STEP 1
0150                LET S=S+T(I,J)
0160                LET Z=Z+T(I,J)
0170            NEXT J
0175 REM
0180            FOR J=1 TO M STEP 1
0190                LET P(I,J)=T(I,J)/S
0200            NEXT J
0210        NEXT I
0214 REM
0215 REM ** COMPUTE COLUMN TOTALS
0216 REM
0220        FOR J=1 TO M STEP 1
0230            FOR I=1 TO M STEP 1
0240                LET C(J)=C(J)+T(I,J)
0250            NEXT I
0260            LET C(J)=C(J)/Z
0270        NEXT J
0274 REM
0275 REM ** COMPUTE AND PRINT CHI-SQUARE AND DF
0276 REM
0280        LET L=0
0290        FOR I=1 TO M STEP 1
0300            FOR J=1 TO M STEP 1
0310                LET L=L+T(I,J)*LOG(P(I,J)/C(J))
0320            NEXT J
0330        NEXT I
```

```
0335 REM
0340          LET L=L*2
0350          PRINT
0360          PRINT
0370          PRINT
0380          PRINT "CHI-SQUARE=  ";L;" WITH "; (M=1)^2;" DF"
0384 REM
0385 REM ** DATA STATEMENTS OF TRANSITION FREQUENCIES
0386 REM
0391          DATA 58,18,2
0392          DATA 15,86,39
0393          DATA 5,35,51
0415 REM
0420      END

END OF LISTING
```

MARK3: TESTING THE STATIONARITY OF A MARKOV CHAIN

```
0005 REM ** TEST OF STATIONARITY OF MARKOV CHAIN
0006 REM ** WRITTEN BY J.N.R.JEFFERS, INSTITUTE OF TERRESTRIAL ECOLOGY
0007 REM ** WRITTEN AUGUST 1982
0008 REM
0009 REM ** THE PROGRAM TESTS THE STATIONARITY OF A SERIES OF MARKOV
0010 REM ** TRANSITION SUB-MATRICES IN EITHER TIME OR SPACE. THE TEST
0011 REM ** IS DISTRIBUTED LIKE CHI-SQUARE, AND TEST VALUES GREATER THAN
0012 REM ** THOSE TABULATED FOR THE GIVEN DEGREES OF FREEDOM INDICATE
0013 REM ** NON-STATIONARITY.
0017 REM
0018 REM ** DIMENSION STATEMENT DEPENDS ON NUMBER OF STATES
0019 REM
0020          DIM N(6,6),P(6,6),Q(6,6)
0024 REM
0025 REM ** INPUT NO OF STATES AND SUB-MATRICES
0026 REM
0030          DISP "NO OF STATES";
0040          INPUT M
0050          DISP "NO OF SUBMATRICES";
0060          INPUT T
0064 REM
0065 REM ** READ AND SUM SUB-MATRICES
0066 REM
0070          MAT Q=ZER(M,M)
0080          FOR I=1 TO T STEP 1
0090              MAT READ N(M,M)
0100              MAT Q=Q+N
0110          NEXT I
0114 REM
0115 REM ** COMPUTE AND PRINT TRANSITION MATRIX
0116 REM
```

```
0120        FOR I=1 TO M STEP 1
0130            LET R=0
0140            FOR J=1 TO M STEP 1
0150                LET R=R+Q(I,J)
0160            NEXT J
0165 REM
0170            FOR J=1 TO M STEP 1
0180                LET Q(I,J)=Q(I,J)/R
0190            NEXT J
0200        NEXT I
0205 REM
0210        PRINT "MATRIX OF TRANSITION PROBABILITIES"
0220        PRINT
0230        PRINT
0240        MAT PRINT Q
0250        PRINT
0260        PRINT
0264 REM
0265 REM ** COMPUTE AND PRINT TEST STATISTIC
0266 REM
0270        RESTORE
0275 REM
0280        FOR K=1 TO T STEP 1
0290            LET S=0
0300            MAT READ N(M,M)
0310            FOR I=1 TO M STEP 1
0315 REM
0320                LET R=0
0330                FOR J=1 TO M STEP 1
0340                    LET R=R+N(I,J)
0350                NEXT J
0355 REM
0360                FOR J=1 TO M STEP 1
0370                    LET P(I,J)=N(I,J)/R
0380                NEXT J
0385 REM
0390                FOR J=1 TO M STEP 1
0400                    LET S=S+N(I,J)*LOG(P(I,J)/Q(I,J))
0410                NEXT J
0415 REM
0420            NEXT I
0425 REM
0430        NEXT K
0435 REM
0440        LET C=2*S
0450        LET D=(T-1)*M*(M-1)
0460        PRINT "CHI-SQUARE = ";C
0470        PRINT "WITH          ";D;"DEGREES OF FREEDOM"
0480        PRINT
0490        PRINT
0500        PRINT "IF THIS CALCULATED VALUE EXCEEDS THE TABULATED VALUE"
```

```
0510          PRINT "OF CHI-SQUARE FOR THE GIVEN NO OF DEGREES OF FREEDOM"
0520          PRINT "THE HYPOTHESIS OF STATIONARITY SHOULD BE REJECTED"
0525 REM
0530 REM ** SUCCESSIVE SUBMATRICES OF FREQUENCIES SHOULD BE"
0540 REM ** GIVEN IN THE DATA STATEMENTS WHICH FOLLOW
0545 REM
0601          DATA
0602          DATA
0603          DATA
0604          DATA
0895 REM
0900       END

END OF LISTING
```

MARK4: COMPUTING SUCCESSIVE MARKOV PROBABILITIES

```
0005 REM ** PROGRAM TO COMPUTE MARKOV PROBABILITIES AFTER N STEPS
0006 REM ** WRITTEN BY J.N.R.JEFFERS, INSTITUTE OF TERRESTRIAL ECOLOGY
0007 REM ** WRITTEN MAY 1979, ANNOTATED JANUARY 1982
0008 REM
0009 REM ** PROGRAM CALCULATES THE TRANSITION PROBABILITIES OF A MARKOV
0010 REM ** PROCESS AFTER N SUCCESSIVE TIME STEPS, PRINTING THE ORIGINAL
0011 REM ** MATRIX OF TRANSITION PROBABILITIES AND THE PROBABILITIES AFTER
0012 REM ** N TIME STEPS HAVE TAKEN PLACE. THE INTERMEDIATE PROBABILITIES
0013 REM ** ARE NOT PRINTED, IN CONTRAST TO MARK3.
0017 REM
0018 REM ** DIMENSION STATEMENT DEPENDS ON NUMBER OF STATES
0019 REM
0020          DIM T(6,6),P(6,6),R(6,6)
0024 REM
0025 REM ** INPUT NO OF STATES AND NO OF TIME STEPS
0026 REM
0030          DISP "NO OF STATES";
0040          INPUT M
0050          DISP "NO OF STEPS";
0060          INPUT N
0064 REM
0065 REM ** READ ORIGINAL TRANSITION MATRIX AS DATA AND PRINT
0066 REM
0070          PRINT "ORIGINAL MATRIX"
0080          PRINT
0085 REM
0090          FOR I=1 TO M STEP 1
0100              FOR J=1 TO M STEP 1
0110                  READ T(I,J)
0120                  LET P(I,J)=T(I,J)
0130                  PRINT T(I,J),
0140              NEXT J
0150              PRINT
0160          NEXT I
```

```
0164 REM
0165 REM ** CALCULATE SUCCESSIVE PROBABILITIES
0166 REM
0170        FOR L=1 TO N STEP 1
0175 REM
0180            FOR I=1 TO M STEP 1
0190                FOR J=1 TO M STEP 1
0200                    LET R(I,J)=0
0210                    FOR K=1 TO M STEP 1
0220                        LET R(I,J)=R(I,J)+T(I,K)*P(K,J)
0230                    NEXT K
0240              NEXT J
0250            NEXT I
0255 REM
0260            FOR I=1 TO M STEP 1
0270                FOR J=1 TO M STEP 1
0280                    LET P(I,J)=R(I,J)
0290                NEXT J
0300            NEXT I
0305 REM
0310        NEXT L
0314 REM
0315 REM ** PRINT FINAL TRANSITION PROBABILITIES
0316 REM
0320        PRINT
0330        PRINT
0340        PRINT "STEP";L-1
0345 REM
0350        FOR I=1 TO M STEP 1
0360            FOR J=1 TO M STEP 1
0370                PRINT P(I,J),
0380            NEXT J
0390            PRINT
0400        NEXT I
0404 REM
0405 REM ** DATA STATEMENTS
0406 REM
0411        DATA 0.77,0.18,0.05
0412        DATA 0.08,0.65,0.27
0413        DATA 0.07,0.32,0.61
0495 REM
0500     END

END OF LISTING
```

AMARK: CALCULATING THE PROPERTIES OF AN ABSORBING MARKOV CHAIN

```
0005 REM ** PROGRAM TO COMPUTE PROPERTIES OF AN ABSORBING MARKOV CHAIN
0006 REM ** WRITTEN BY J.N.R.JEFFERS, INSTITUTE OF TERRESTRIAL ECOLOGY
0007 REM ** WRITTEN OCTOBER 1978 FROM AN ALGORITHM BY J.G.KEMENY
```

```
0008 REM ** ANNOTATED JANUARY 1982
0009 REM
0010 REM ** PROGRAM CALCULATES THE BASIC PROPERTIES OF A MARKOV TRANSITION
0011 REM ** MATRIX WITH ONE OR MORE ABSORBING STATES. IN ADDITION TO THE
0012 REM ** FUNDAMENTAL MATRIX, THE PROBABILITIES TO ABSORPTION FROM
0013 REM ** EACH STATE ARE CALCULATED AND PRINTED
0014 REM
0018 REM ** DIMENSIONS DEPEND ON NUMBER OF STATES
0019 REM
0020        DIM Q(10,10),N(10,10),R(10,10)
0030        DIM T(10),U(10)
0034 REM
0035 REM ** READ NO OF ABSORBING AND NO OF TRANSIENT STATES
0036 REM
0040        READ M,N
0044 REM
0045 REM ** READ MATRIX OF TRANSIENT PROBABILITIES
0046 REM
0050        MAT READ Q(N,N)
0054 REM
0055 REM ** READ MATRIX OF TRANSIENT TO ABSORBING PROBABILITIES
0056 REM
0060        MAT READ R(N,M)
0064 REM
0065 REM ** PREPARE IDENTITY MATRIX AND VECTOR
0066 REM
0070        MAT N=IDN(N,N)
0080        MAT U=CON(N,1)
0084 REM
0085 REM ** PRINT TRANSIENT TO TRANSIENT PROBABILITIES
0086 REM
0090        PRINT "TRANSIENT TO TRANSIENT PROBABILITIES"
0100        MAT PRINT Q
0110        PRINT
0114 REM
0115 REM ** PRINT TRANSIENT TO ABSORBING PROBABILITIES
0116 REM
0120        PRINT "TRANSIENT TO ABSORBING PROBABILITIES"
0130        MAT PRINT R
0140        PRINT
0144 REM
0145 REM ** COMPUTE AND PRINT FUNDAMENTAL MATRIX
0146 REM
0150        MAT Q=N-Q
0160        MAT N=INV(Q)
0170        PRINT "FUNDAMENTAL MATRIX"
0180        MAT PRINT N
0190        PRINT
0200        MAT T=N*U
0210        MAT PRINT T
0220        PRINT
```

```
0224 REM
0225 REM ** COMPUTE AND PRINT ABSORPTION PROBABILITIES
0226 REM
0230        MAT Q=N*R
0240        PRINT "ABSORPTION PROBABILITIES"
0250        MAT PRINT Q
0694 REM
0695 REM ** DATA HELD AS DATA STATEMENTS
0705 REM
0795 REM ** MATRIX OF TRANSIENT PROBABILITIES
0796 REM
0801        DATA 0.936, 0.064, 0.000, 0.000
0802        DATA 0.000, 0.921, 0.079, 0.000
0803        DATA 0.000, 0.000, 0.951, 0.049
0804        DATA 0.000, 0.000, 0.000, 0.961
0844 REM
0845 REM ** MATRIX OF ABSORBING PROBABILITIES
0846 REM
0850        DATA 0.000, 0.000, 0.000, 0.039
0895 REM
0900     END

END OF LISTING
```

EMARK: CALCULATING THE PROPERTIES OF AN ERGODIC MARKOV CHAIN

```
0005 REM ** PROGRAM TO CALCULATE THE PROPERTIES OF AN ERGODIC MARKOV CHAIN
0006 REM ** WRITTEN BY J.N.R.JEFFERS, INSTITUTE OF TERRESTRIAL ECOLOGY
0007 REM ** WRITTEN AUGUST 1980 FROM AN ALGORITHM BY J.G. KEMENY
0008 REM ** ANNOTATED FEBRUARY 1982
0009 REM
0010 REM ** PROGRAM CALCULATES THE BASIC PROPERTIES OF AN ERGODIC MARKOV
0011 REM ** CHAIN. IN ADDITION TO THE LIMITING PROBABILITIES FOR EACH STATE
0012 REM ** THE PROGRAM CALCULATES THE MEAN FIRST PASSAGE TIMES AND THE
0013 REM ** FIRST PASSAGE TIMES IN EQUILIBRIUM.
0014 REM
0018 REM
0019 REM
0020        DIM P(10,10),M(10,10),Z(10,10)
0030        DIM A(1,10),B(1,10)
0034 REM
0035 REM ** READ NO OF STATES AND MATRIX OF TRANSITION PROBABILITIES
0036 REM
0040        READ N
0050        MAT READ P(N,N)
0054 REM
0055 REM ** CONSTRUCT IDENTITY MATRIX
0056 REM
0060        MAT Z=IDN(N,N)
0064 REM
```

```
0065 REM ** PRINT MATRIX OF TRANSITION PROBABILITIES
0066 REM
0070        PRINT "TRANSITION PROBABILITIES"
0080        PRINT
0090        MAT PRINT P
0100        PRINT
0110        PRINT
0120        PRINT
0124 REM
0125 REM ** COMPUTE AND PRINT LIMITING PROBABILITIES
0126 REM
0130        MAT Z=Z-P
0135 REM
0140        FOR I=1 TO N STEP 1
0150           LET Z(I,N)=1
0160        NEXT I
0165 REM
0170        MAT M=INV(Z)
0180        MAT A=ZER(1,N)
0185 REM
0190        FOR J=1 TO N STEP 1
0200           LET A(1,J)=M(N,J)
0210        NEXT J
0215 REM
0220        PRINT "LIMITING PROBABILITIES"
0230        PRINT
0240        MAT PRINT A
0250        PRINT
0260        PRINT
0270        PRINT
0274 REM
0275 REM ** COMPUTE AND PRINT MEAN FIRST PASSAGE TIMES
0276 REM
0280        MAT M=IDN(N,N)
0290        MAT M=M-P
0295 REM
0300        FOR I=1 TO N STEP 1
0310           FOR J=1 TO N STEP 1
0320              LET M(I,J)=M(I,J)+A(1,J)
0330           NEXT J
0340        NEXT I
0345 REM
0350        MAT Z=INV(M)
0355 REM
0360        FOR I=1 TO N STEP 1
0370           FOR J=1 TO N STEP 1
0380              LET M(I,J)=(Z(J,J)-Z(I,J))/A(1,J)
0390           NEXT J
0400        NEXT I
0405 REM
0410        PRINT "MEAN FIRST PASSAGE TIMES"
```

```
0420          PRINT
0430          MAT PRINT M
0440          PRINT
0450          PRINT
0460          PRINT
0464 REM
0465 REM ** COMPUTE AND PRINT FIRST PASSAGE TIMES IN EQUILIBRIUM
0466 REM
0470          MAT B=A*M
0475 REM
0480          PRINT "FIRST PASSAGE TIMES IN EQUILIBRIUM"
0485          PRINT
0490          MAT PRINT B
0494 REM
0495 REM ** DATA HELD AS DATA STATEMENTS
0496 REM
0497 REM ** NUMBER OF STATES
0498 REM
0500          DATA 4
0504 REM
0505 REM ** MATRIX OF TRANSITION PROBABILITIES
0506 REM
0510          DATA 0.38, 0.18, 0.00, 0.44
0520          DATA 0.45, 0.32, 0.21, 0.02
0530          DATA 0.31, 0.04, 0.17, 0.48
0540          DATA 0.40, 0.16, 0.22, 0.21
0595 REM
0600       END

END OF LISTING
```

Index